Listeria monocytogenes

Edited by **Monde Alfred Nyila**

Listeria monocytogenes
http://dx.doi.org/10.5772/intechopen.71807
Edited by Monde Alfred Nyila

Contributors

Frederick Tawi Tabit, Cristina Saraiva, Nihed Ben Halima, Prasann Kumar, Shweta Pathak, Paula Teixeira, Vânia Ferreira, Sofia Pereira, Ângela Alves, Elena Karanina, Ekaterina Selezneva, Monde Alfred Nyila

First published in London, United Kingdom, 2018 by IntechOpen
IntechOpen is the global imprint of INTECHOPEN LIMITED, registered in England and Wales, registration number: 11086078, The Shard, 25th floor, 32 London Bridge Street
London, SE19SG – United Kingdom
Printed in Croatia

British Library Cataloguing-in-Publication Data
A catalogue record for this book is available from the British Library

Additional hard copies can be obtained from orders@intechopen.com

Listeria monocytogenes, Edited by Monde Alfred Nyila
p. cm.
Print ISBN 978-1-78923-628-6
Online ISBN 978-1-78923-629-3

For EU product safety concerns:
IN TECH d.o.o., Prolaz Marije Krucifikse Kozulić 3, 51000 Rijeka, Croatia,
info@intechopen.com or visit our website at intechopen.com.

Meet the editor

Dr. Monde A. Nyila graduated with his PhD degree in Plant Science from the University of Pretoria, South Africa; his MSc degree in Environmental Biotechnology (2001) from the Rhodes University, South Africa; and his BSc degree (Honours) in Microbiology (1991) from UWC, South Africa. His PhD thesis was titled "Antilisterial Bioactivity and/or Biofilm-Formation by Compounds from *Plectranthus ecklonii* Benth. and *Acacia karroo* Hayne". The research was on medicinal plants that have antibacterial activity against the foodborne pathogen, *Listeria monocytogenes*. He has presented papers at local and international conferences. He has coauthored articles on *Listeria monocytogenes* in peer-reviewed journals. He is a senior lecturer and teaches microbiology and phytomedicine at the University of South Africa. He is involved in the supervision of MSc and PhD students. He has also coauthored a book chapter published by IntechOpen.

Contents

Preface

The quest to ensure human safety with regard to the prevalence of foodborne diseases caused by foodborne pathogens such as *Listeria monocytogenes* necessitates the prevention and the awareness of factors that contribute to poor healthcare delivery systems. In order to provide a cost effective healthcare system with regard to illnesses caused by foodborne pathogens, the healthcare workers should know how to approach the disease by coming up with healthcare programs that would limit the spread of the disease.

Since prevention is better than cure, the methods involved in food storage, handling, preparation, and serving could save many lives. The preventative methods could save governments and health institutions large amounts of money. The treatment of listeriosis involves the use of antibiotics, which can sometimes lead to drug resistance or multidrug resistance (MDR).

The aim of this publication is to provide well-researched up-to-date information on the foodborne pathogen, *Listeria monocytogenes*, to the readers, health practitioners, policy makers, food industry workers, researchers, and all stakeholders involved in food safety and food security.

I would like to thank and congratulate all the contributing authors for their professionalism, hard work, and patience in making this book a great success. Last but not least, I would like to thank Mr. Markus Mattila, the Author Service Manager, for his role in the completion of this book project.

Dr. Monde A. Nyila
University of South Africa
Republic of South Africa

Section 1

Ubiquity of Listeria monocytogenes

Chapter 1

Introductory Chapter: *Listeria monocytogenes*

Monde A. Nyila

Additional information is available at the end of the chapter

http://dx.doi.org/10.5772/intechopen.76905

1. Introduction

The purpose of this book is to give informative well-researched chapters to the readers, health practitioners, policy makers, food industry, researchers and all stakeholders involved in food safety and food security. The chapters in this book are from specialists in their respective disciplines. The unprecedented outbreak of listeriosis in South Africa recently from January 2017 to March 2018 [1] has made this book to be more important. This has been reported as the greatest outbreak of listeriosis in recent times because of the number of cases reported as well as the number of fatalities.

The book is about the foodborne pathogen, *Listeria monocytogenes*. The pathogen is of public health concern [2], which causes serious diseases such as endocarditis, encephalitis sepsis and meningitis, gastro-enteritis and death [3, 4]. Infected pregnant women suffer from flu-like febrile illness, and depending on the stage of pregnancy, it may lead to abortion. Listeriosis is a notifiable disease in most industrialised countries with few or no report from Africa. It became a notifiable disease recently in South Africa after the unprecedented outbreak. In countries where listeriosis is a reportable disease, the countries have come up with actions that keep surveillance of food and food-processing plants [5].

L. monocytogenes is a ubiquitous pathogen because it is found everywhere. It has been reported [6] that the municipal wastewater effluent was a source of listerial pathogen in the aquatic environment. Surveillance in developing countries is on assumption that the cases of listeriosis are not always reported [7] and that traditional methods used for food storage eliminate means of growth of microorganisms. The outbreak of listeriosis has been widely reported in Japan, North America and Europe [8].

2. Why it is important to know much about *L. monocytogenes*?

Many lives would be saved if more research on *L. monocytogenes* can be done as well as educating the public about hygiene, food handling, preparation and distribution and cooking. The declaration of the listeriosis outbreak was done in December 2017 in South Africa [1] although the listeriosis outbreak occurred long before December 2017. The reported cases in South Africa were as follows: 743 cases in 2017 and 202 cases in the first 2 months of 2018. In the reported cases, female accounted for 55% and neonates aged ≤28 days accounted for 41% [1]. Gauteng Province had the most reported cases of 59% followed by the Western Cape with 17% and KwaZulu-Natal with 7%. The National Health Laboratory Services (NHLS) sampled over 1500 food stuffs from retail outlets, food-processing plants and patients. The molecular sequences were done at the NICD. Over 70 items tested positive for *L. monocytogenes*. **Figure 1** taken from NICD shows the confirmation of listerial infection per age distribution. The ready-to-eat meat products, which include Russians, ham, other 'cold' meats, sausages, Viennas and Polonies, were found to be the source of listerial outbreak.

Figure 1 shows that neonates were severely affected by the outbreak of listeriosis in South Africa. The results are similar to the reported [8] maternofetal listeriosis or neonatal listeriosis which presented life-threatening illness. Gastrointestinal listeriosis has been widely reported in the developed countries as shown in **Table 1**.

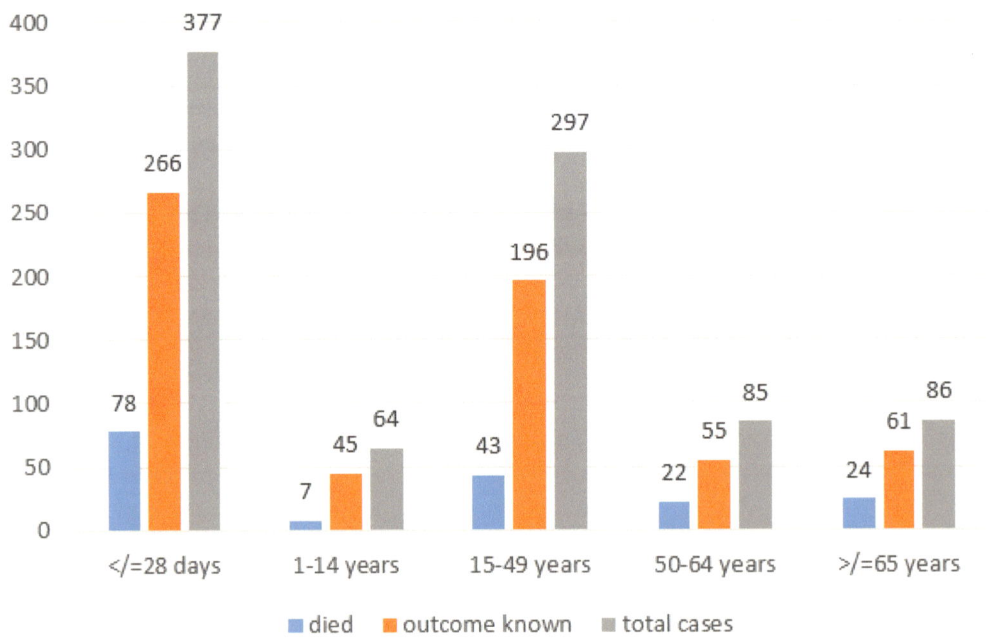

Figure 1. Age distribution and outcome of laboratory-confirmed cases of listeriosis identified from 1 January 2017 to 27 February 2018 (n = 909 where age was reported) [1].

Year	Location	Number of cases	Implicated source
1993	Northern Italy	18	Rice salad
1994	IL, USA	44	Chocolate milk
1997	Northern Italy	1566	Cold corn and tuna salad
1998	Finland	N/A	Cold-smoked fish
2000	New Zealand	32	Ready-to-eat meat
2001	CA, USA	16	Delicatessen turkey ready-to-eat meat
2001	Sweden	48	Raw milk cheese
2001	Japan	38	Cheese

N/A, number of cases not given.

Table 1. Gastrointestinal listeriosis outbreaks, 1993–2001 (adapted from [8]).

3. Chapters in this book

The chapters in this book cover a vast scope with regard to the *L. monocytogenes* pathogen.

The topics cover *L. monocytogenes* in medical research, quality assurance of raw food material, virulence traits of *L. monocytogenes* relevant to food safety, and so on. The topics investigate at *L. monocytogenes* in all angles such as pathogenicity, virulence, stress factors, susceptibility, prevention and control.

Acknowledgements

I would like to thank the Publishing Process Manager, Markus Mattila, for his assistance in communicating and liaising with the contributing authors and also keeping me up to speed with regard to the submissions of chapter proposals and full chapters. I also thank my institution, the University of South Africa, for giving me support.

Author details

Monde A. Nyila

Address all correspondence to: nyilama@unisa.ac.za

Department of Life and Consumer Sciences, University of South Africa, Johannesburg, Florida, Republic of South Africa

References

[1] NICD. Situation Report on Listeriosis Outbreak, South Africa. 2018. http://www.nicd.ac.za/wp-content/uploads/2018/02/Listeria-Sitrep-27-Feb-2018.pdf [Accessed on March 3, 2018]

[2] Schmid B, Klumpp J, Raimann E, Loessner MJ. Role of cold shock proteins in growth of *Listeria monocytogenes* under cold and osmotic conditions. Applied and Environmental Microbiology. 2009;**75**:1621-1627

[3] De Souza VM, Franceschini SA, Martinez RCR, Ratti RP, De Martinis ECP. Survey of Listeria spp. in matched clinical, food and refrigerator samples at home level in Brazil. Food Control. 2008;**19**:1011-1013

[4] Goldenberg RI, Thompson C. Infectious origins of stillbirth. American Journal of Obstetrics and Gynecology. 2003;**189**:861-873

[5] Pagotto F, Ng L-K, Clark C, Farber J. Canadian Listeriosis reference service. Foodborne Pathogens and Disease. 2006;**3**:132-137

[6] Odjadjare EE, Obi LC, Okoh AI. Municipal wastewater effluents as a source of listerial pathogens in the aquatic milieu of the Eastern Cape Province of South Africa: A concern of public health importance. International Journal of Environmental Research and Public Health. 2010;**7**(5):2376-2394

[7] Todd ECD, Notermans S. Surveillance of listeriosis and its causative pathogen, Listeria monocytogenes. Food Control. 2011;**22**:1484-1490

[8] Swaminathan B, Gerner-Smidt P. The epidemiology of human listeriosis. Microbes and Infection. 2007;**9**:1236-1243

Section 2

Listeriosis, Ready-to-eat Food, Food Processing and Food Safety

Chapter 2

Quality Assurance of Food Raw Materials and Food Products as the Main Factor of Safety of the Consumer Market

Elena V. Karanina and Ekaterina Y. Selezneva

Additional information is available at the end of the chapter

http://dx.doi.org/10.5772/intechopen.75407

"You are what you eat"

–Hippocrates.

Abstract

Ensuring food safety in the consumer market of the country, region, city is one of the priority activities to improve the quality of life of the population. To ensure the quality and safety of food products in the consumer market of the country needs consistent work. Necessary implementation of measures on formation of culture of healthy eating, food security, improving the nutritional status of the population and protection of health. Most businesses manufacture and sell products that comply with the requirements of the current normative-technical documentation, although this documentation, the experts have questions. But despite this, in the consumer market there is low-quality, non-standard and adulterated food products. One of the most pressing issues of concern to the public authorities, manufacturers, retailers, public organizations and, of course, the customers, is the prevention of falsification and counterfeiting. When exercising the state control over observance of requirements of technical regulations on food products throughout the Russian Federation the complex of measures aimed at curbing the production and trafficking of counterfeit food products. The focus is on the dairy industry, childcare, healthcare organizations, trade enterprises. And require further tightening, so as to completely solve the problem of falsification and counterfeiting is not impossible to solve. In order to prevent receipt of the consumer market of the country, the region, poor-quality food raw materials and food products necessary to intensify the work of public associations (manufacturers, distributor, wholesalers), to ensure the publicity of product quality from manufacturers, to achieve a reliable assessment of the quality of their products by manufacturers or Association.

Keywords: the quality and safety of food raw materials and food products, the safety of the consumer market, violations, falsification of products, counterfeit

1. Introduction

In the past 30 years the overall incidence of the Russian population is constantly growing, due, on the one hand, the increase in the elderly population and better detection of diseases through new methods of diagnosis, and with another—deterioration of the health of the population and the inefficiency of the system for the prevention and treatment of diseases. Food can be the source and carrier of significant quantities dangerous to human health of toxic substances. The rapid growth of production, environmental problems, expanding the range of products and unfair attitude of the producer to what he produces has led to the fact that the quality of food has decreased significantly. These factors relate to priority socio-economic problems that generate negative trends in health status and mortality.

Unauthorized use in the process of agricultural production of medicinal products for veterinary use, intentionally introduced into an organism of productive animals, leading to contamination of food and to negative consequences for human health (appearance of infectious agents with new properties, increasing the severity and consequences of infection, antibiotic resistance, allergic reactions), requiring increased costs of their treatment, including the provision of high-tech medical care [4].

Ensuring food safety in the consumer market of the country, region, city is one of the priority activities to improve the quality of life of the population. To ensure the quality and safety of food products in the consumer market of the country requires consistent work.

2. Methodology

Theoretical and methodological basis of the research was the concepts and hypotheses presented in the works of domestic and foreign scientists on safety, food security, regional economy, regional consumer markets, research on the quality of goods.

The methodological substantiation of the research on the chosen topic is formed on the principles of the system approach and the general theory of systems, the theory of quality systems, the theory of logistics and trade management. In order to substantiate the propositions put forward in the study and solve these problems, methods of economic-mathematical modeling, system analysis, structural analysis, statistical, expert-analytical, sociological methods were used, which allowed revealing the essence of the issues studied, obtaining reliable results and providing sufficient justification for the estimates and conclusions contained in the study.

3. Results of the study

In the structure of causes of death in the Kirov region as in the whole of the Russian Federation (according to the operative data of Rosstat) still account for the bulk of diseases of the circulatory system (48.5%), neoplasms (15.7%), accidents, poisoning and traffic injuries (9.4%).

Mortality from external causes of death exceed the average value is 1.3 times, from diseases of the circulatory system and neoplasms 1.2 times. Despite a steady trend of annual decline of mortality from accidental alcohol poisoning in 2016 this figure in the region continues to exceed the national average by 3.3 times.

In the structure of mortality from external causes in the Kirov region attract the attention of high mortality rates from suicide (above the average for Russia in two times) (**Table 1**).

In dynamics in recent years, still maintains a positive downward trend in mortality for all major causes of death except for deaths from neoplasms in the last 3 years there has been a rise in mortality from this cause. In recent years the level of general morbidity with the diagnosis established for the first time has a tendency to decrease (**Table 2**). The incidence rate in 2016 below the average annual value of 3%. Compared to 2012, the year of the primary incidence of the total population of the region in 2016 decreased by 4.3% [5].

The structure of the first identified incidence of the region's population in 2015 has not changed significantly. The most common cause of primary morbidity of population of the region, as in previous years, were diseases of the respiratory system. Second place in the morbidity structure of children and adults is injury, poisoning and certain other causes external causes, and among teenagers for the first time in many years, in second place came eye disease (**Table 3**).

Analysis of the average annual rate of growth (decline) of the disease has enabled to identify classes of diseases characterized by a tendency to increase. So, in 2016, a tendency to increase (compared to 2014) are the primary indicators of child morbidity diseases of the endocrine system (1.2 times) due to, primarily, the growth of obesity among children (1.3 times), the marked increase of congenital anomalies (malformations) in children (1.2 times) [3].

The main causes of death	**2012 year**	**2013 year**	**2014 year**	**2015 year**	**2016 year**	**Russian Federation (2016 year)**
Deaths from all causes:	1560.3	1536.0	1513.5	1518.6	1490.1	1288.3
From some infectious and parasitic diseases	8.8	9.5	9.6	9.1	8.2	22.3
Tumors	211.0	214.3	233.2	225.7	234.7	201.6
Diseases of the circulatory system	937.5	888.2	770.3	807.5	722.4	614.1
Diseases of the respiratory system	70.6	72.1	76.4	66.1	51.5	47.1
Diseases of the digestive system	59.5	57.3	66.8	72.9	71.8	66.6
External causes of death:	188.6	175.1	170.0	154.6	140.2	104.8
Of them from traffic injuries	20.7	19.6	19.3	16.0	14.3	14.7
Alcohol poisoning	36.0	24.9	29.1	23.4	19.0	5.7
Suicide	37.3	38.6	36.1	33.9	31.9	15.6
Murders	12.2	12.8	10.6	9.7	9.2	7.0

Table 1. Mortality of the population of the Kirov region by main causes of death in 2012–2016 g (per 100,000 population) [4].

Indicators	2012 year	2013 year	2014 year	2015 year	2016 year	Russian Federation (2016 year)
All diseases from them:	788.5	767.2	788.6	755.0	754.8	778.2
Certain infectious and parasitic	35.1	33.7	33.3	33.3	28.3	28.1
Tumors	8.9	8.8	9.5	9.9	9.9	11.4
Blood and blood-forming organs and certain disorders involving the immune mechanism	4.6	4.7	4.6	4.5	5.3	4.7
Endocrine, nutritional and metabolic disorders	10.1	11.2	10.4	10.7	17.4	13.3
Nervous system	12.0	12.1	12.3	11.0	11.3	15.4
Eye and adnexa	31.5	30.9	31.9	31.9	32.0	33.3
Ear and mastoid	27.0	27.7	26.5	24.5	25.3	26.6
Circulatory	23.2	22.5	31.2	26.3	29.8	31.2
Respiratory	372.6	349.7	372.7	349.0	349.8	337.9
Digestive	18.4	17.3	16.8	18.6	18.0	35.3
Skin and subcutaneous tissue	40.6	42.9	44.1	41.2	36.8	44.0
Musculoskeletal system and connective tissue	28.7	28.0	24.8	23.2	25	30.1
Genitourinary system	33.6	33.3	34.0	35.1	32.6	46.4
Congenital anomalies (malformations), deformations and chromosomal abnormalities	1.0	1.1	0.8	0.8	1.0	2.0
Injury, poisoning and certain other consequences of external causes	99.8	101.9	97.0	98.5	95.4	90.4

Table 2. Incidence of the Kirov region population by main classes of diseases (patients registered with the diagnosis set for the first time, per 1000 of the population) [4].

Rank	Kids	Teens	Adults
First place	Diseases of the respiratory system was 72.4%	Diseases of the respiratory system is 57.5%	Diseases of the respiratory system–25.4%
Second place	Injury, poisoning and certain other consequences of external causes is 5.3%	diseases of the eye and adnexa–7.5%	Injuries, poisoning and some other consequences of external causes by 18.7%
Third place	Infectious, parasitic diseases–4.0%	Injuries, poisoning and some other consequences of external causes 6.3%	Of illnesses of system of blood circulation–7.4%
Fourth place	diseases of the eye and adnexa–2.6%	Of the disease customizing system is 5.1%	Diseases of the genitourinary system–7.0%
Fifth place	Diseases of the skin and subcutaneous tissue at 2.5%	Diseases of the skin and subcutaneous tissue is 4.8%	Diseases of the skin and subcutaneous tissue–6.6%
Sixth place	Sixth place diseases of the ear and mastoid–2.4%	Diseases of the genitourinary system and 3.7%	Diseases of the eye and adnexa–5.3%

Table 3. The structure of primary morbidity of the population of the Kirov region in 2016 [4].

In the adult population is significant statistically significant increased incidence in 2015 compared to previous year is observed for diseases of the endocrine system (1.8 times), including obesity in 2.4 times, thyroid disease 1.4 times, for diseases of the blood (1.6-fold), diseases of the circulatory system (19.5%).

Among the sanitary-hygienic risk factors of human health, the main contribution of complex chemical load due to the contamination of drinking water, air, food, soil.

Ensuring food safety in the consumer market of the country, region, city is one of the priority activities to improve the quality of life of the population. To ensure the quality and safety of food products in the consumer market of the country needs consistent work.

Necessary implementation of measures on formation of culture of healthy eating, food security, improving the nutritional status of the population and protection of health.

Most businesses manufacture and sell products that comply with the requirements of the current normative-technical documentation, although this documentation, the experts have questions. But despite this, in the consumer market there is low-quality, non-standard and adulterated food products.

Food security of the population is one of the most important socio-economic tasks, as the degree of satisfaction of food affects human health, their well-being. The development of the food market depends on the state of agro-industrial complex as the main source of formation of resource base in the trading sector, the solvency of the buyers and consumer demand.

The food market is adversely affected by the absence of a developed market infrastructure, high cost levels in production, discrepancy of products to the required standards. In General, the food market of Russia and Kirov region in particular in recent years there has been positive dynamics of development, his state was stable in almost all types of products. The food market area was characterized by saturation, absence of physical scarcity, high competition on the market of food products, increasing demand, growth in retail trade turnover [2].

The concept of "food security" in relation to the regions is not sufficiently well-established in contrast to the definitions used at the country level and opened in the food security Doctrine of the Russian Federation. According to the Doctrine of "food safety of Russia–is the state of the economy, which ensures food independence of the Russian Federation, is guaranteed physical and economic accessibility for every citizen of the country of food products that meet the requirements of the legislation of the Russian Federation on technical regulation, in amounts not less than the rational norms of food consumption necessary for an active and healthy lifestyle."

Ensuring the safety of food raw materials and food products is one of the main areas that determine the health of the population and its genetic conservation. With foods in the human body comes 40–50% of harmful substances, water 20–40. In connection with the above, you must submit analysis of the incidence of one of the regions of the Russian Federation, namely in Kirov region. Monitoring of drinking water quality identified areas of risk where the population uses drinking water that does not meet sanitary requirements on sanitary-to chemical indicators. Prolonged use of drinking water with high levels of contamination by chemicals of natural and anthropogenic character could be one of the reasons for the development of various non-communicable diseases in the population [6].

The toxic effect of nitrates is associated with restoring them to nitrites, ammonia, hydroxylamine, under the influence of microorganisms and enzymes of the digestive tract. It is the nitrites may have an adverse effect on people, as a direct (through the formation of methemoglobin) and indirect (through synthesis of carcinogenic compounds–nitrosamines).

Priority contaminants of food products on the territory of the Kirov region are nitrate, unsatisfactory research results, the contents of which are recorded ieoeie in fruits and vegetables.

You must also submit an analysis of officially identified substandard goods and counterfeit in Russia (**Table 4**).

In General, it can be noted that the level of quality increases according to official data, but still there are cases of detecting substandard, counterfeit products (**Figure 1**). According to a study by the Russian Institute of consumer tests (RIPI) food held in 2016 identified the following violations [9].

Results of the study of food products in 2016 are as follows. Does not meet safety requirements:

- 48% of vegetable production;
- 38% of fish products;

Food supply	2013 year		2014 year		2015 year		2016 year	
	Domestic	Imported	Domestic	Imported	Domestic	Imported	Domestic	Imported
Meat and poultry	13	6	2	4	5	3	3	3
Poultry meat	4	10	3	8	7	2	3	4
Sausage goods	2	0	1	1	3	2	2	1
Production fish food commodity (without fish canned food)	6	3	4	18	8	19	2	1
Whole milk products	1	3	4	2	6	3	1	3
Pasta	0	0	1	0	1	0	0	10
Cereal	1	0	1	11	1	1	1	3
Flour	2	0	0	0	1	0	1	0
Pastry	4	27	2	25	4	6	2	1
Butter	2	0	4	2	4	0	4	1
Vegetable oil	0	0	0	2	0	0	0	0
Production of margarine and mayonnaise	1	0	3	2	2	0	0	2
Cheeses	1	4	3	4	2	5	2	1

Table 4. Quality of goods received by the consumer market, Russian Federation, percentage of the number of samples [10].

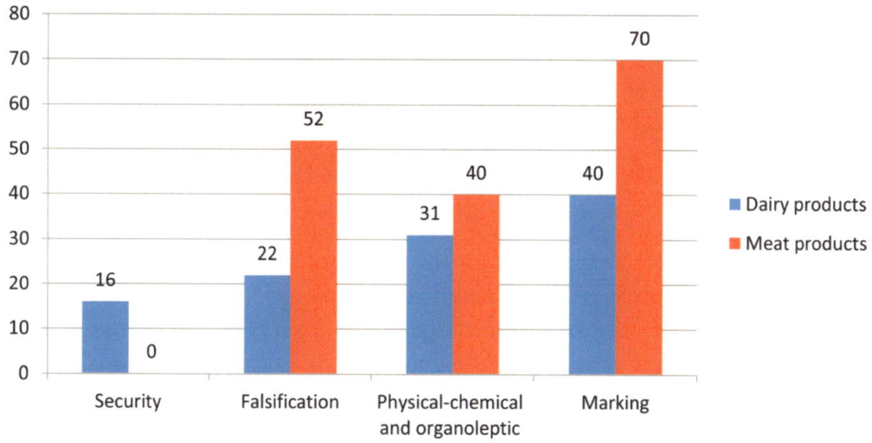

Figure 1. Chart of violations in a separate food group in 2016 (%).

- 24% of groceries;
- 16% of dairy products [10].

A high percentage of dangerous vegetable products is due to the presence in the fresh vegetables of residual amounts of pesticides that are not regulated to mandatory testing (only 2 pesticides are to be checked when declaring the product) (**Figure 2**). Is falsified: 52% of meat products; 22% of dairy products.

The specificity of the revealed falsification of meat products is due to the fact that the actual composition of the products differs from the one declared on the packaging (replacement of raw materials). Also, manufacturers use additives to improve the final product, using poor quality raw materials. Histological analysis of the composition of the product makes it possible to identify the presence of undeclared components and/or the absence of declared ones. When using

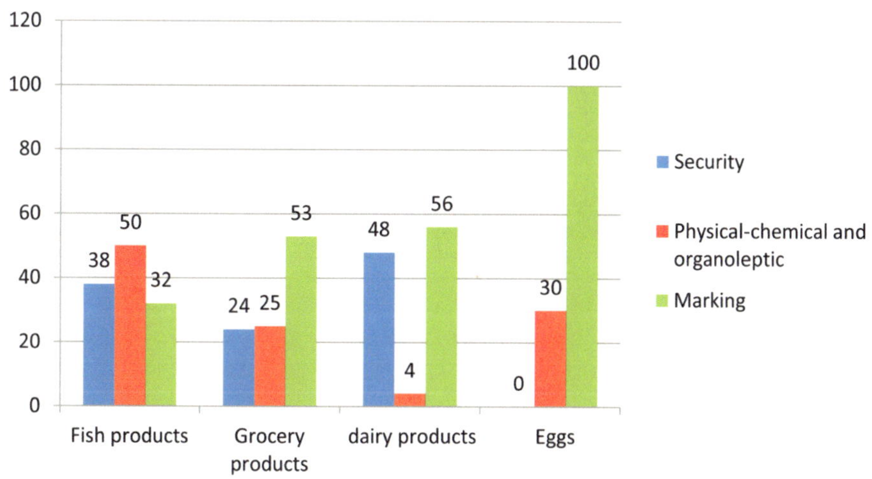

Figure 2. Chart of violations in a separate food group in 2016 (%).

low-grade meat or raw materials with an increased moisture content (frozen meat), appropriate water absorbers or water-retaining additives (starch, flour, carrageenan, etc.) are used that transfer excess moisture to a colloidal or emulsion state. Another option of falsification is the addition of artificial flavors, dyes and other food additives, including preservatives.

Among the meat falsifications there are traditional leaders. These are such popular products as sausage "Doctor's" (68%) and sausages (50%), produced in accordance with GOST, and pelmeni (64%).

In the category of dairy products, the proportion of counterfeit products is 22%, excluding the RIPI studies for three groups of dairy products: packaged curd, cream and ghee. In them, the results of laboratory analyzes do not allow to make an unambiguous conclusion that the product is not falsified; **Table 5**) .

Food safety research conducted in 2016 revealed the following:

- The situation is extremely alarming in the vegetable market–almost every second sample is dangerous (48%), while the share of dangerous fresh vegetables and melons is even higher (60%).
- As in previous years, the safety of fish products is a concern–every third sample is dangerous (38%).
- The share of dangerous dairy products was at least 16%. Compared to 2015, it decreased by more than a third, but it should be noted that safety indicators in dairy products were tested in less than half of the studies.
- The share of dangerous groceries was not less than 24% (safety indicators were checked in less than half the studies).

Research has revealed a large share of falsified products: every second meat product (52%) is falsified, and at least one in five milk products (22%). It should be noted that the existing methods of quality control of dairy products do not allow to unambiguously and reliably qualify products as falsified. According to physicochemical indicators and organoleptic, the share of products with violations of technical regulations is as follows: fish—50%, meat products—40%, dairy—31%, bread—21%. With respect to labeling, it can be noted that consumers' rights to complete and reliable information the product is violated systematically.

Falsifiers in dairy products	Does not comply with the physicochemical parameters and organoleptics	Share of violations on marking
Sour cream–33%;	50% of fish products;	100%–eggs
Butter–30%;	40% of meat products;	70%–meat products
Cottage cheese–22%;	31% of dairy products;	56%–fresh vegetables
Milk–21%;	30% of eggs;	53%–groceries
Ice cream plombir–20%;	25% of groceries.	40%–dairy products
Condensed milk–17%;		32%–fish products
Cheese–4%.		

Table 5. Data on violations by groups.

The revealed infringements (on safety, falsification and marking of products) testify to non-compliance with the requirements of the technical regulations of the Customs Union, the RF Law "On Protection of Consumer Rights" and other legislative acts [7].

It is also necessary to present the experience of identifying falsifications and substandard products in the Kirov region in 2017. The facts of turnover of meat and meat products have been established with violation of storage conditions, sale of meat without the presence of safety documents, lack of necessary labeling for meat products, inadequate sanitary maintenance of the territory, violation of personal hygiene and disinfection regulations, lack of organization and conduct of industrial laboratory control. In the course of inspections, 52 batches of meat products were rejected and withdrawn from sale. Similar violations have been identified for fruits and vegetables, cape and fish and other products. For example, in the framework of the state control of the quality and safety of food raw materials and food products on the territory of the Kirov region, samples of fat and oil products were selected. The indicated products for the fatty acid composition of milk fat and the ratio of the mass fractions of methyl esters of fatty acids in milk fat do not meet the requirements. This indicates the falsification of the fatty phase of the oil with fats of non-dairy origin.

Low-quality level of quality is also noted by consumers of products. The survey conducted by the author in 2016 revealed the following problems of retail chains in the Kirov region, according to consumers (**Figure 3**).

Thus, among the main problems typical for the lion's share of the Kirov retailers, respondents most often mentioned violations of quality standards (38%), unsatisfactory sanitary condition (21%) and a rise in prices for goods (16%). In aggregate, more than half of the respondents note the importance of the quality factor and their dissatisfaction with the goods and services sold by retail chains.

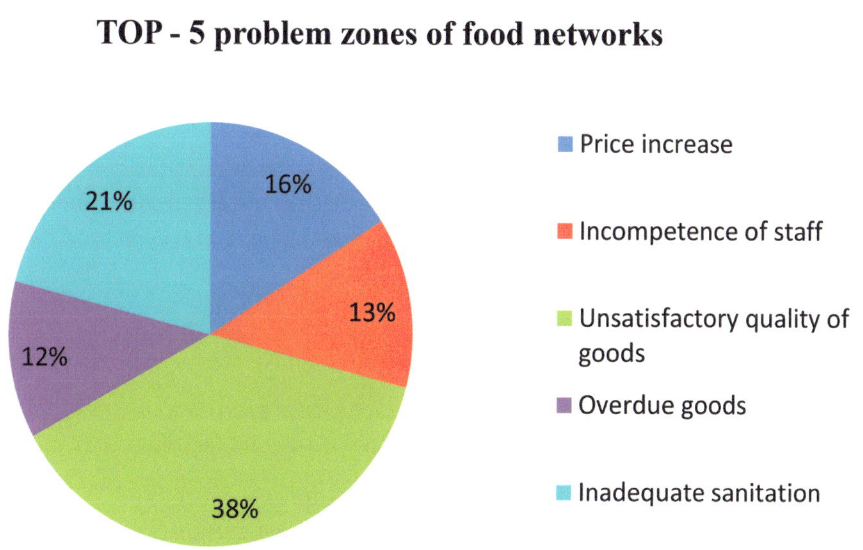

Figure 3. The opinion of the Kirov consumers about the problems in retail chains.

Speaking about food safety, it is necessary first of all to raise the issue of environmentally friendly raw materials for their production. This problem must be solved both at the state level and in the regions. Until recently, restrictions on the content of harmful substances were shown only to the final product–food products–and did not apply to the raw materials from which they were produced [8].

The unsatisfactory quality of food products, namely, the high content of nitrates, pesticides, mycotoxins, and heavy metals in them not only leads to a decrease in their consumer properties (the content of vitamins and essential amino acids decreases, the composition of macro- and microelements changes, and the organoleptic properties also decrease products), but, more importantly, to negative effects on the human body as a whole. Thus, the consumption of food products with a high content of nitrates and nitrites leads to an increase in the content of methemoglobin in the blood, which leads to a decrease in the supply of oxygen to the organs and tissues of the body and subsequent disruption of the functions of all body systems, including the activity of the central nervous and cardiovascular systems [7].

Accumulation in the body of pesticides leads to the development of severe variants of gastroenterological diseases with frequent prolonged exacerbations associated with damage to the mucous membrane of the gastrointestinal tract, as well as the concomitant development of neurovegetative, endocrine, immunological, dysbiotic and functional cardiovascular disorders. And the accumulation in the human body of heavy metals and their salts, leads to damage to absolutely all systems of the body, including the central nervous, cardiovascular, respiratory, digestive and endocrine systems, as well as to disorders of the musculoskeletal system.

The unsatisfactory situation in the food market of the Russian Federation and the Kirov region in particular can be explained by the following factors:

1. Control over the process of production and sale of food products is withdrawn from the sphere of state regulation and left to the discretion of producers. Also, the safety of food products in Russia is regulated by about 7000 standard documents, but none of them affects the quality of products.

2. Many Russian food industry enterprises do not always create conditions in which product safety is possible. It is practically impossible to realize, in the absence of a modern system of quality control and safety of food raw materials and finished types of food products.

3. Another significant problem is that food products are often perishable, as a result of physiological processes and microbiological effects.

4. It should also be noted that the retail network is seriously affecting the level of product quality, and for the worse. For example, manufacturers are forced to increase the shelf life, reduce prices for products, all this is achieved only through deterioration in quality.

Undoubtedly, we need a new certification system for compliance with ISO 22000: 2005 (GOST R ISO 22000–2007) "Food safety management system. Requirements for the organization of the chain of production and supply." But this measure will not solve the problem. A national quality system is required. Currently, a national food quality management system is being developed. Rosselkhoznadzor and Roskazhestvo should coordinate actions aimed at dynamically increasing the level of quality of domestic products [3].

The system of traceability of raw materials and commodity batches created by Rosselkhoznadzor allows to control the whole chain of production and distribution of agricultural raw materials and food products, as well as to identify and suppress risks associated with their inferiority and falsification. Rosselkhoznadzor will inform Roskazhestvo about the state of safety of food products sold in Russia. In turn, Roskazhestvo will provide the Rosselkhoznadzor with the results of its own fan research. The tracking system will come into effect already this winter. The new system of interaction will allow: to quickly identify risks of contamination and falsification of food products; make the turnover of products in the domestic market more transparent; ensure timely detection and withdrawal from the trading network of a low-quality batch of goods.

But despite the positive developments, it should be noted that the system of regulatory and legal regulation of relations in the field of quality assurance developed in the Russian Federation is based on the norms of the World Trade Organization and the Eurasian Economic Union.

Imperfection of legal and organizational mechanisms with respect to the quality of food products leads to the appearance of low-quality foreign goods and frank falsifications on the Russian market [5].

Consumption of food products with low consumer properties is the reason for the decline in the quality of life and the development of a number of diseases of the population, including due to unreasonably high caloric content of food, reduced nutritional value, excessive intake of saturated fats, micronutrient deficiency and dietary fiber. What is further exacerbated by the lack of a unified information system for quality control of food products throughout all processes of food production and circulation, to track the use of medicines for veterinary use and plant protection products, to identify the organizations responsible for each stage in the chain of its production and circulation.

The problem of ensuring the quality of food products is also the almost complete absence in the Russian Federation of the production of food ingredients and substances (vitamins, amino acids, food additives, enzyme preparations, biologically active substances, starter and probiotic microorganisms, prebiotic substances, etc.).

The existing system of methods for controlling both food additives and food additives in the composition of food products requires improvement.

In order to improve the quality and safety of food products, a number of regulatory documents have been adopted in the Russian Federation, such as the "Food Security Doctrine of the Russian Federation" and "The Strategy for Improving the Quality of Food Products until 2030". The adoption of the "Food Security Doctrine" is aimed at reliable provision of the country's population with safe agricultural products, fish and other products from aquatic biological resources and foodstuffs, which in turn should lead to an improvement in the quality of life of Russian citizens by guaranteeing high standards of livelihood [11].

Within the framework of the "Strategy for improving the quality of food products until 2030" adopted on July 4, 2016, it is planned to improve the regulatory framework and product monitoring systems. A unified information system should be created, through which consumers will be able to obtain data on the composition of a product and its manufacturer. The strategy is aimed at providing adequate nutrition, preventing diseases, increasing the duration and improving the quality of life of the population, stimulating the development of production and circulation in the food market of proper quality.

Within the framework of solving the problem of the quality of consumer goods and raw materials, it is necessary to solve a number:

- Improvement and development of the regulatory framework in the field of food quality, including legal aspects related to effective compensation mechanisms for the protection of consumers' rights;
- improving and developing the methodological basis for assessing the compliance of food quality indicators;
- ensuring the monitoring of the quality of food products;
- Improvement of state regulation in the field of food quality, including in terms of ensuring state control (supervision) and applying administrative penalties for non-compliance by the manufacturer (executor, seller, a person performing the functions of a foreign manufacturer) with food quality requirements;
- creation of a single information system for the traceability of food products;
- development and implementation of a quality management system for food products;
- creation of incentive mechanisms for producers to produce food products that meet the quality criteria and principles of healthy eating;
- creation of conditions for the production of new generation food products with specified quality characteristics;
- the revival of the production of food ingredients in the Russian Federation;
- actualization of the current standards for the maintenance of food additives, flavors, biologically active substances, remnants of medicinal products for veterinary use and plant protection products in food products;
- priority development of scientific research in the field of nutrition of the population, including in the field of prevention of the most common non-communicable diseases and development of production technologies aimed at improving the quality of food products;
- Promoting the principles of healthy eating;
- Close work with retail chains to improve the quality of food at the federal and regional levels, using a policy of "national" and "regional" protectionism.

The strategy is the basis for the formation of a national food quality management system [9]. Thus, the need to improve the quality, as well as the development of functional foods that can become part of the daily diet, and can have a positive impact in preventing the development of various diseases of all systems of the human body and reduce the rate of premature aging of the organism due to external factors is topical. This task becomes especially important in the tense ecological situation in the Sverdlovsk region, taking into account the ever increasing negative impact of such sanitary and hygienic risk factors as chemical, biological and radiation dose loads on public health.

4. The conclusion

Based on the analysis, the following conclusions can be drawn. Previous studies did not have a systematic analysis of the relationship between the health of the population of the region and the nutrition of the population. In the structure of the causes of death in the Kirov region, the major part is still the diseases of the circulatory system, neoplasms, accidents, poisoning and transport injuries.

In the adult population, there is a significant increase in the incidence rate in 2016 for endocrine system diseases, including obesity, thyroid disease, blood diseases, circulatory system diseases. Among the sanitary and hygienic risk factors of health disorders, the main contribution is made by the complex chemical load due to contamination of drinking water, atmospheric air, food, soil. It is necessary to implement measures to create a healthy diet, ensure food security, improve the quality of nutrition of the population and protect its health.

Food provision of the population is one of the most important socio-economic tasks, since the degree of satisfaction with food products depends on a person's health and well-being.

Currently, does not meet the safety requirements: 48% of vegetable products, 38% of fish products, 24% of groceries, 16% of dairy products. Research has revealed a large share of falsified products: every second meat product (52%) is falsified, and at least one in five milk products (22%). According to physicochemical indicators and organoleptic, the share of products with violations of technical regulations is as follows: fish—50%, meat products—40%, dairy—31%, bread—21%. With respect to labeling, it can be noted that consumers' rights to complete and reliable information the product is violated systematically.

Among the main problems typical for the lion's share of the Kirov retailers, respondents most often mentioned breaches of quality standards (38%), unsatisfactory sanitary condition (21%) and a rise in prices for goods (16%).

With the purpose of solving the problem of the quality of consumer goods and raw materials, it is necessary to solve a number of tasks: improving and developing the regulatory framework in the field of food quality, improving and developing the methodological base, ensuring food quality monitoring, improving state regulation in the field of food quality, actualization of existing standards for the content of food additives in food, priority development of scientific research in the nutrition of the population, promoting the principles of healthy eating, working closely with retailers. These measures will improve the quality of products and consequently increase the level of public health and reduce the level of mortality.

Author details

Elena V. Karanina* and Ekaterina Y. Selezneva

*Address all correspondence to: kafinanc@yandex.ru

Vyatka State University, Kirov, Russia

References

[1] Presidential Decree On Approving the Doctrine of Food Security of the Russian Federation. Official website [Electronic resource]: URL: http://www.kremlin.ru/events/president/news/6752 (reference date: January 24, 2013)

[2] Order of the Government of the Russian Federation on the approval of the "Strategy for improving the quality of food products in the Russian Federation until 2030". Official website [Electronic resource]: URL: http://government.ru/docs/23606/ (date of application: 01-24-2018)

[3] On signing an agreement on information cooperation between the Rosselkhoznadzor and Roskaschestvo [Electronic resource]. Rosselkhoznadzor/Official site [Electronic resource]: URL: http://www.fsvps.ru/fsvps/print/news/15581.html (date of treatment: 01-23-2018)

[4] On the state of sanitary and epidemiological welfare of the population in the Kirov region in 2016: State report-Office of the Federal Service for Supervision of Consumer Rights Protection and Human Welfare in the Kirov Region; 2017. p. 198

[5] Gadzhieva SR, Alieva TI, Abdullaev RA, Velieva ZT. Problems of food safety. Young Scientist. 2014;**4**:417-418. URL https://moluch.ru/archive/63/9425/ (reference date: January 14 2013)

[6] Baykhozhaeva BU. Quality and safety of food products: issues and problems [Electronic resource]. Eurasian National University. Gumilev LN [Electronic resource]: URL: http://repository.enu.kz/bitstream/handle/123456789/4487/kachestvo_bezopasnost.pdf (date of circulation: 01-21-2018)

[7] Valievich RP. Quality Management of Goods and Services: A Manual. Vol. 11. Minsk: BSEU; 2013

[8] Garina EP, Garin AP. Use of modern quality management methods to improve the production performance of a complex product. Vestnik NGIER. 2017;**11**(78):111-119

[9] Molnar P. Modern problems in the field of quality management and food safety in Europe [Electronic resource]. Website about quality management [Electronic resource]: URL: http://quality.eup.ru/MATERIALY14/eur_mk_bp.htm (date of circulation: January 22, 2013)

[10] Kupriyanov AV. The System of Ensuring the Quality and Safety of Food Products. Bulletin of the Orenburg State University. 2014;**3**(164):164-167

[11] Kupriyanova LM. Product quality: Problems and solutions. The World of a New Economy. 2015:375-85

[12] Petrusevich TV. Features of the analysis of the quality of functioning of non-commercial organizations. Economics. Business. Banks. Ns 1. pp. 34-42

[13] Sosunova IA. Health of modern man: Environmental aspects. Electronic periodic scientific publication Bulletin of the International Academy of Sciences. Russian Section. 2014;**1**:43-46

[14] Tsinke T, Kimatova RG, Kubasheva GA. Enterprise quality management: analysis and models of quality management. 2017;**12**(3):87-95

[15] Federal Service of State Statistics Electron. text dan. URL: http://www.gks.ru (reference date: 01/21/2018)

[16] Territorial body of the Federal State Statistics Service for the Kirov region: officer.site. Electron. text dan. URL: http://www.kirovreg.ru/econom/industry/import.php (date of circulation: 01/21/2018)

[17] Vyatka commercial and industrial fees: officers.site]. Electron. text dan. URL: http://www.vcci.ru (date of circulation: 01/21/2018)

[18] The official website of the government of the Kirov region. URL: http://www.kirovreg.ru/econom/itog.php (reference date: January 14, 2018)

Chapter 3

Modeling the Behavior of *Listeria monocytogenes* in Meat

Cristina Saraiva, Juan García-Díez,
Maria da Conceição Fontes and Alexandra Esteves

Additional information is available at the end of the chapter

http://dx.doi.org/10.5772/intechopen.79967

Abstract

This work was conducted to present some studies that show the behavior of *Listeria monocytogenes* in meat, according to intrinsic and extrinsic factors. The understanding of factors that affect the survival and growth of *L. monocytogenes* in meat, such as temperature, pH, acid, salt, water activity or modified atmosphere packaging, is crucial to develop strategies for food operators to reduce and prevent *Listeria* contamination and growth. The knowledge of *L. monocytogenes* behavior according to its physiological and ecological characteristics, under all probable conditions, will support risk assessors to find strategies to control this ubiquitous bacteria in food industry and food service. The Regulation (EC) No 2073/2005, reviewed by the Regulation (EC) No 1441/2007, does not establish the limits for *L. monocytogenes* in fresh meat. However, it is generally accepted a level of 100 cells on fresh meat, except for some risk groups. Food business operators and authorities can use predictive microbiology models as important tools to model bacterial growth in quantitative microbial risk assessments.

Keywords: *Listeria monocytogenes*, meat, modified atmosphere packaging (MAP), antimicrobial agents, predictive microbiology

1. Introduction

Meat is a protein food commodity with a significant water content that makes it a great matrix susceptible to bacterial growth [1]. Since meat forms part of the dietary habits of consumers, several strategies to improve its safety, shelf-life and quality have been studied in the recent years.

The genus *Listeria* includes a group of Gram-positive psychrotrophic bacteria that can be isolated from a large variety of environmental sources such as water, soil, foodstuffs, animals or humans [2, 3]. Also, *Listeria* can colonize various inert surfaces (e.g., surfaces of food machinery) [4]. Genus *Listeria* includes nonsporulating, catalase positive, Voges-Proskauer positive, indol and oxidase negative, facultative anaerobic rods that show motility at 25°C. *Listeria* can also grow in a large variety of conditions like high salt concentrations, low water activity, broad pH range (pH 4.5–9) and broad range of temperature (0–45°C, optimum 30–37°C) [3, 5]. This genus *Listeria* includes several species such as *L. monocytogenes, L. ivanovii, L. innocua, L. welshimeri, L. seeligeri, L. grayi, L. marthii, L. rocourtiae, L. leichmannii, L. weihenstephanensis, L. floridensis, L. aquatica, L. cornellensis, L. riparia and L. grandensis* [6]. Among them, *L. monocytogenes* is the most important due to its pathogenicity that affects animals and humans. The ingestion of contaminated foods is the most important source of human infection. According to somatic (O) and flagellar (H) antigens, 13 serotypes of *L. monocytogenes* have been recognized and identified alphanumerically as 1/2a, 1/2b, 1/2c, 3a, 3b, 3c, 4a, 4ab, 4b, 4c, 4d, 4e and 7. Serotypes 1/2a, 1/2b and 1/2c are the most frequently isolated from both foodstuffs and food processing areas [6, 7]. Concerning the different kind of implicated food vehicles of listeriosis outbreaks, cooled meats, ready-to-eat foods, cheese, smoked fish and seafood seem to be more susceptible to *L. monocytogenes* development [8–10]. According to [9], in 2016, a total of 0.47 cases of listeriosis per 100,000 population was reported with an incidence of about 10% compared to the previous year. In addition, mortality achieved 16% among the confirmed cases. For most healthy people, listeriosis does not imply more than a threat, limited to gastrointestinal symptomatology ended in 36–48 hours. However, life-threatening infections mainly occur in high-risk populations, including pregnant women, neonates, infants, elderly and individuals with compromised immune systems [10]. Clinical features of listeriosis have considerable variability and can be confused with other infections. Sometimes, gastrointestinal manifestations as primary infection are observed. These digestive manifestations are usually self-limited and spontaneously resolved [11]. *L. monocytogenes* has a tropism for the central nervous system causing meningitis. Sepsis without a localized infection is the most common presentation in patients with deficient immune systems [12]. Since *L. monocytogenes* is a microorganism of ubiquitous nature, meat and meat products may become contaminated throughout contact with raw materials, processing environment and at retail markets [4], *L. monocytogenes* can adhere to the surfaces forming biofilms [13], which consist of cells and extra-cellular polymeric materials that protect bacteria and lead to its survival and growth. Indeed, *L. monocytogenes* showed some resistance against biocides [13] and temperature-dependent resistance to phages. Therefore, finding alternative methodologies to avoid the contamination and further survival and growth of *L. monocytogenes* are important requests of meat industry [14].

The Regulation (EC) No 2073/2005, reviewed by the Regulation (EC) No 1441/2007, does not establish the limits for *L. monocytogenes* in fresh meat. However, a level of 100 cells on fresh meat, except for some risk groups, is generally accepted.

This chapter provides a brief background on *L. monocytogenes* as an important foodborne pathogen and describes the main factors, such as temperature, pH, acid, salt, water activity or packaging, that influence its behavior in meats. There will be referred some control strategies for control of survival and growth of *L. monocytogenes* and the advantageous use of predictive microbiology programs.

2. Factors related to survival and growth of *L. monocytogenes* in meat

2.1. Influence of pH

The optimal pH growth of *L. monocytogenes* is between 6 and 8 [15]. However, *L. monocytogenes* can adapt, grow and survive in acid environments. Its resistance depends on other ecological factors and its physiological condition. The influence of the pH in the growth of *L. monocytogenes* was largely studied [1, 16]. With reference to [1], it was shown that ultimate pH (normal and DFD meats) influenced significantly the growth of *L. monocytogenes* inoculated on beef samples stored at two temperatures (4 and 9°C). The growth of *L. monocytogenes* was higher on DFD meat, revealing the effect of the ultimate pH with evident dependence of the storage temperature. In fact, at 4°C, no growth of *L. monocytogenes* was observed on meat with normal ultimate pH, but on DFD meat, this bacteria achieved levels of 5.5 log CFU/g in vacuum-packed samples at day 14 of storage. At this time, levels of 8 log CFU/g was obtained in vacuum-packed meats stored at abusive temperature (9° C) [1].

It has been shown that tolerance to low pH can be induced in *L. monocytogenes* by exposure to sublethal pH conditions. During the adaptation period, *L. monocytogenes* synthesizes a set of proteins that allows it to survive under stress conditions. Thus, according to [17], it was evidenced the existence of proteins, using two-dimensional gel electrophoresis, which are only present in acid stress conditions. The activation of several genes responsible for the codification of the proteins that confer resistance in pH stress conditions has been discussed by several authors [18, 19]. The stress sigma factor (σ^B) has been referred as responsible for *L. monocytogenes* resistance, although other genes are also involved in the resistance mechanisms [19, 20].

The resistance of *L. monocytogenes* to acid conditions may compromise the safety of several foodstuffs with low pH. It should be taken into consideration especially in foodstuffs with a long shelf-life in which survival microorganisms might be associated to an outbreak.

2.2. Osmotic stress

The resistance of *L. monocytogenes* at low aW values depends on environmental factors as well as its physiological condition. Most of the reports assessed the osmotic resistance together with other factors such as temperature or pH. [21] reported that minimum aW value required for *L. monocytogenes* growth generally increased with the reduction of temperature. The range of aW of *L. monocytogenes* that allows growth is variable. According to [22], it was showed that aW resistance depends on the initial counts of *L. monocytogenes* in broth. Thus, at low contamination (between 1 and 20 CFU/ml), an inhibition of *L. monocytogenes* was observed at aW values above 0.975. However, the critical aW for *L. monocytogenes* growth was set at 0.965 at high contamination levels (between 500 and 1000 CFU/ml). With reference to [23], the authors inoculated 10^7 CFU/ml in Mueller-Hinton broth at three levels of aW (0.91, 0.95 and 0.97) to study the effect of aW factor. Although a reduction, of about 1 log CFU/ml, was observed after incubation for 4 hours at 0.97 of aW, *L. monocytogenes* survivors developed after 24 hours of *lag* phase, showing that can adapt to the osmotic stress condition. *L. monocytogenes* is tolerant to NaCl, and it was capable to grow in 25.5% and survived for 1 year in 16% NaCl [24].

The adaption of *L. monocytogenes* to osmotic stress is associated to three main mechanisms: induction of proteins, accumulation of solutes as osmoprotectants and the stress sigma factor. The induction of proteins has been observed by [25] throughout electrophoresis analysis. Although the mechanisms are not clearly understood, some genetic interference has been reported by [26]. The accumulation of osmoprotectants to maintain the osmotic balance such as glycine betaine, proline betaine, acetyl carnitine or carnitine was described by [27]. The amount of each accumulated osmolyte by the cell appeared to be dependent of the growth media osmolarity. All of them play an important role in the osmoprotection, although with reference to [28] observed that carnitine is not as effective as glycine betaine in contributing to either salt or chill stress responses of *L. monocytogenes*. The stress sigma factor is induced upon exposure of *Listeria* to several stress conditions and improves the resistance of *Listeria* by regulation the production of protective substances [29].

2.3. Temperature

L. monocytogenes is capable to survive and to multiply over a wide range of temperatures. The lower limit for the growth of *L. monocytogenes* in food matrices with a high content of nutrients and neutral pH is around 0°C. With reference to [1], the final storage of air and vacuum-packed beef samples stored at abusive temperatures (9°C) produced higher (2–3 log CFU/g) counts of *L. monocytogenes* than observed in samples stored at 4°C. The presence of *L. monocytogenes* in refrigerated meat products during the product shelf-life has been reported by [30]. Although refrigeration is a common conservation method, the indiscriminate use of cold, that is, in sliced dry-cured meat products [23] may improve the survival of *L. monocytogenes* [31] reported that *L. monocytogenes* has grown in air-packed beef stored at 5°C up to 16 days. The *lag* phase is variable according to the environment temperature and may be associated to the physiological modification of *Listeria* to survive at low temperatures. The changes in the membrane composition at low temperatures may lead to a change in the membrane lipid composition in order to maintain the fluidity required for proper enzymatic activity and solute transport. Growth at low temperatures also results in an increase in the percentage of unsaturated fatty acids to improve the membrane fluidity [32].

Low temperatures lead to changes in gene expression and induction of proteins named cold shock proteins in response to temperature shocks. In consequence, this adaptation of *Listeria* implies changes in its gene expression [33]. As previously discussed, the accumulation of solutes such as glycine, betaine and carnitine acts as a cryoprotectant. Moreover, the role of the alternative sigma factor B (σ^B) is associated to the resistance of *L. monocytogenes* at low temperatures as it may be involved in the stimulation of the genes responsible for the synthesis and accumulation of the cryoprotectant solutes [34].

2.4. Packaging

The growth of *L. monocytogenes* is scarcely affected by anaerobic or oxygen-reduced atmosphere. According to [16, 35], modified atmosphere packaging (MAP) systems may reduce the survival and growth of *L. monocytogenes* by the presence of carbon dioxide in modified atmosphere packaging (MAP). In fresh beef, MAP with 60% CO_2: 30% O_2: 10% N_2 prevented growth at 4°C for more than 2 weeks of storage. Although regarding vacuum packaging, this preservation methodology seems not affecting the growth of *L. monocytogenes* as observed by [16]. With reference to [35], it was showed that *L. monocytogenes* survives better in vacuum packaging than in air-packed beef samples. According to [36], neither *L. monocytogenes* grow after 42 days of storage nor significant reductions were observed in inoculated vacuum-packed beef stored at 4°C [37] observed *L. monocytogenes* growth in vacuum-packaged beef stored at 0 and 5°C. They indicated that growth of this bacteria on beef depends on the storage temperature, pH and the type of tissue (fat or lean). Although *L. monocytogenes* grows at both temperatures, a scarce *lag* period was observed in beef stored at 5°C. Similarly, [35] observed an increase of *lag* phase of *L. monocytogenes* in beef samples stored at 4°C compared to those stored at 9°C. In a study with pork cuts [38] stored at mean refrigerator temperatures did not increase the populations of *L. monocytogenes* over 2 log CFU/g in the end of product shelf-life. However, at abusive temperatures, microbial counts were higher than 3 log CFU/g for some cases, which required a more severe heat inactivation treatment before consumption. According to the *lag* phase of *L. monocytogenes* in vacuum-packed beef at 0°C, it was extended until 60 days. Regarding the type of the tissue, a faster growth of *L. monocytogenes* was observed in fat than in lean that may be associated to the differences of pH of both tissues. In consequence, to improve the beef safety against *L. monocytogenes*, the storage at low temperatures and vacuum packaging must be associated to other barrier systems such as bacteriocins or essential oils [39].

3. Strategies for *L. monocytogenes* growth control in meat

Classically, the main methodologies for fresh meat preservation are chilling and freezing, but technologies such as packaging systems like modified atmosphere packaging (MAP) and active packaging (AP) or use of natural antimicrobial compounds have arisen to improve its safety and quality.

Currently, consumers' growing concern about chemical hazard in foods reflects an increased awareness about the harmful effects that they may have on human health. In consequence, the consumers' demand on more healthy and natural foods, leading food industry to use natural substances such as plant extracts, essential oils, chitosan and organic acids to satisfy this green consumering tendency.

The use of essential oils (EOs) to control *L. monocytogenes* has been studied by several authors. With reference to [40], the antimicrobial effect of thyme EO against *L. monocytogenes* in minced beef during 12 days of storage at 4°C was studied, and about 2 log CFU/g reduction of *L. monocytogenes* counts was observed after 2 days of storage. Although an increase of *L. monocytogenes* counts was found after 6 days of storage, indicating a potential of its adaptation to the EO. Similar results were observed by [41] in minced meat inoculated with thymus EO, although concentrations of thymus EO at 0.25 and 1.25% decreased progressively the counts of *L. monocytogenes* up to 15 days of storage at 7°C.

In a study of [39] rosemary EO sprayed in beef samples presented a greater inhibitory effect against *L. monocytogenes* compared to thymus EO was reported. This fact can be related to the chemical composition of this EO since the concentration of phenolic compounds (i.e., thymol) was lower than the obtained in rosemary EO. With reference to [42], an antimicrobial effect of oregano, cinnamomum, rosmarinus, salvia and thymus EO against *L. monocytogenes* in meatballs stored at 4°C was observed, while the extension of the antimicrobial effect varied according to the added EO and its concentration. A reduction of 1 and 2 log CFU/g, on average, was observed when concentrations of about 1 and 2% were added, respectively. It indicates that the antimicrobial effect of EO in foodstuffs is not enough to guarantee the safety of meat in case of high contamination. In addition, the negative impact on sensory acceptance was also indicated by the authors.

Regarding active packaging, the addiction of several substances with antimicrobial effect such as organic acids, chitosan or nisin among others has also been studied to improve meat safety against *L. monocytogenes*. With reference to [43], the authors observed that the use of chitosan diluted in acetic acid or lactic acid as coating in highly contaminated ready-to-eat roast beef (6.5 log/CFU) is useful to control *L. monocytogenes*. The use of sodium lactate and sodium diacetate in edible coating in combination with polysaccharide-based edible coatings have been studied by [44] in chilled and frozen roasted turkey. Although organic acids decreased the counts of *L. monocytogenes*, its combination with chitosan increased the antimicrobial effect. With reference to [45] who evaluated the decontaminating efficacy of lactic acid (2%), potassium sorbate (1%), sodium hypochlorite (200 ppm) and ethanol (10%) sprayed on the surface of meat previously inoculated with 100 μL of a suspension of *L. monocytogenes* (1.5×10^4 CFU/g), the authors observed that samples treated with lactic acid showed significantly lower counts than the controls and other treatments. Lactic acid was shown to be promising in the control of *L. monocytogenes* presenting an early bactericidal effect.

The use of *Lactobacillus sakei* to control *L. monocytogenes* in fresh beef was reported by [46]. Incorporation of lactic acid bacteria into sodium-caseinate films protected beef by lowering the growth of *L. monocytogenes* during storage under abusive temperatures. This strategy could be useful to guarantee the safety of fresh beef along the food chain in which temperature fluctuations may occur.

Bacteriophages harmless to human cells are considered natural biocontrol agents against foodborne pathogens [47]. Bacteriophages are bacterial viruses with host specificity and lysis activities and can be used as preservatives or for pathogens rapid detection [48].

Phages used for biocontrol purposes should be virulent and feature abroad host range, that is, infect and kill as many target strains as possible [49, 50]. Virulent myoviruses closely related to P100 and A511 are the most popular and have been isolated from the sources in Europe, the US and New Zealand [51–53]. Commercially, "ListexTM P100" is available that was generally recognized as safe (GRAS) by FDA and USDA in 2007 for use in all food products. Several studies have showed its efficacy in foods such as ready-to-eat (RTE) meats and poultry [50, 51]. The phage A511, closely related to P100, also showed efficacy in various RTE foods [54]. According to [55], the direct immobilization of the viral particles in the cellulose membranes of the packaging materials can be used in alternative to the phage suspension as a possible intervention strategy against *Listeria*.

4. Predictive microbiology models of *L. monocytogenes*

Predictive microbiology models are used to infer about the evolution of microbial population considering the initial contamination and food environment, as the responses of microorganisms populations in a specific environment are reproducible [56]. Mathematical models may be generally categorized into three types: primary, secondary and tertiary models. The primary models are used to estimate the changes in the microbial population as a function of time, under a single set of conditions [57, 58]. The secondary models describe the microorganisms' responses to environmental conditions, according to one or more parameters of a primary model [59]. The tertiary models were defined by [60] as algorithms incorporated into software to integrate the effect of environmental variables on microbial responses and to provide predictions of the outcomes.

The increasing interest in the behavior of hazards such as *L. monocytogenes* promoted important advances in predictive microbiology, and it started to use the food matrix, instead of culture media [61]. Traditional strategies using fast-growing strains in optimal growth conditions usually overestimate the bacterial growth in a food product. This can lead to safe results but may also conduct to unnecessarily safety measures. A stochastic (or probabilistic) approach take into account the variability and uncertainty of various factors that affecting microbial behavior by using probability distributions of the input data. This provides safe enough predictions to avoid unacceptable health risks for consumers [62].

Predictive microbiology models are important tools to model bacterial growth in quantitative microbial risk assessments (MQRA) [63]. In this context, food business operators and authorities can use accessible predictive models, such as Pathogen Modeling Program [64], SymPrevius [65] and ComBase [66]. The incorporation of predictive microbiology models in MQRA must follow some guidelines [56]. The complexity of the predictive microbiology model elected in a MQRA depends on different factors, namely the needs of risk assessment, available model and data availability [63].

For an assessment of microorganisms' behavior in naturally contaminated foods, biological factors, food characteristics and storage conditions must be considered [67]. These authors emphasize the variability of *L. monocytogenes* growth in foods. According to [61], the role of microbial competition in models is now taken into consideration. Some studies were published regarding the survival of *L. monocytogenes* in fresh beef stored at two temperatures and different packaging systems as modified atmosphere packaging (MAP), using omnibus model based on the Weibull Equation [35]. Besides the increase of studies using predictive models, there are few data referred to the application of predictive models to composite foods containing raw and cooked ingredients [67].

According to [63], it is challenging for a risk assessor to choose an applicable predictive microbiology model in the abundant literature. This author suggests that the choice of a model should be done with the closed cooperation between microbiologists, mathematicians and risk managers.

5. Conclusions

Besides the Regulation (EC) No 2073/2005, reviewed by the Regulation (EC) No 1441/2007 does not establish limits for *L. monocytogenes* in fresh meat, it is generally accepted a level of 100 cells on fresh meat, except for high-risk populations. Thus, the implementation of control procedures during processing and at retail level is important. These measures are closely dependent on intrinsic and extrinsic meat factors that could influence microbial growth, namely pH and storage temperature.

Several strategies to improve meat safety and shelf-life have been studied in the latest years. From those, the use of alternative meat packaging systems has been strongly studied to obtain an attractive meat with a higher shelf-life. However, in some cases, these strategies associated to refrigerated storage can promote the survival and growth of some pathogenic microorganisms such as *L. monocytogenes*. However, some authors referred that independently of the refrigeration temperature, the presence of CO_2 in the package atmosphere exerted a bactericidal effect on *L. monocytogenes* cells.

Food business operators and authorities can use predictive microbiology models as important tools to model bacterial survival or growth in quantitative microbial risk assessments. There are several mathematical models to predict the behavior of microorganisms in meat and meat products. However, predictive microbiological models must be carefully used and by whom who is expertise and has an understanding of their limitations and conditions of use.

Acknowledgements

The authors would like to thank CECAV-UTAD. This work was supported by the Portuguese Science and Technology Foundation (FCT) under the UID/CVT/00772/2013 and UID/CVT/00772/2016 projects.

Conflict of interest

The authors declare no conflict of interests.

Author details

Cristina Saraiva*, Juan García-Díez, Maria da Conceição Fontes and Alexandra Esteves

*Address all correspondence to: crisarai@utad.pt

CECAV, Animal and Veterinary Research Centre, Department of Veterinary Sciences, School of Agrarian and Veterinary Sciences, University of Trás-os-Montes e Alto Douro, Vila Real, Portugal

References

[1] Saraiva C, Jardim MC, Fontes MC, Patarata L, Esteves A, Ultimate MC. Packaging on behavior of *Escherichia coli* O157:H7 and *Listeria monocytogenes* inoculated on beef of Maronesa breed at 4 and 9°C. In: Proceedings of the Micro'05-Biotec'05 Congress. Sociedade Portuguesa de Microbiologia e a Sociedade Portuguesa de Biotecnologia. Póvoa de Varzim, Portugal; 2005. p. 202

[2] Prescott M, Harley P, Klein D. Microbial growth. In: Prescott LM, Harley JP, Klein DA, editors. Microbiology. 6th ed. New York: McGraw-Hill Publishing; 2005. pp. 109-132

[3] Barbuddhe S, Hain T, Chakraborty T. The genus listeria. In: Goldman E, Green LH, editors. Practical Handbook of Microbiology. 2nd ed. New York: CRC Press; 2009. pp. 533-562

[4] Norwood DE, Gilmour A. The growth and resistance to sodium hypochlorite of *Listeria monocytogenes* in a steady-state multispecies biofilm. Journal of Applied Microbiology. 2000;**88**:512-520. DOI: 10.1046/j.1365-2672.2000.00990.x

[5] Sauders BD, Wiedmann M. Ecology of *Listeria species* and *L. monocytogenes* in the natural environment. In: Ryser ET, Marth EH, editors. Listeria, Listeriosis, and Food Safety. New York, USA: Marcel Dekker; 2007. pp. 21-53

[6] Meloni D. Focusing on the main morphological and physiological characteristics of the food-borne pathogen *Listeria monocytogenes*. Journal of Veterinary Science and Research. 2014;**1**:1-2

[7] Meloni D. Presence of *Listeria monocytogenes* in Mediterranean-style dry fermented sausages. Food. 2015;**4**:34-50. DOI: 10.3390/foods4010034

[8] Martins EA, Germano PML. *Listeria monocytogenes* in ready-to-eat, sliced, cooked ham and salami products, marketed in the city of São Paulo, Brazil: Occurrence, quantification and serotyping. Food Control. 2011;**22**:297-302. DOI: 10.1016/j.foodcont.2010.07.026

[9] EFSA (European Food Safety Authority), European Centre for Disease Prevention and Control (ECDC). The European Union summary report on trends and sources of zoonosis, zoonotic agents and food-borne outbreaks in 2016. EFSA Journal. 2017;**15**(12):5077

[10] Bover-Cid S, Belletti N, Garriga M, Aymerich T. Model for *Listeria monocytogenes* inactivation on dry-cured ham by high hydrostatic pressure processing. Food Microbiology. 2011; **28**(4):804-809. DOI: 10.1016/j.fm.2010.05.005

[11] Lecuit M. Human listeriosis and animal models. Microbes and Infection. 2007;**9**(9–29): 1216-1225. DOI: 10.1016/j.micinf.2007.05.009

[12] Marquis H, Drevets DA, Bronze MS, Kathariou S, Golos TG, Iruretagoyena JI. Pathogenesis of *Listeria monocytogenes* in humans. In: Singh SK, editor. Human Emerging and Re-Emerging Infections: Viral and Parasitic Infections. Hoboken, New Jersey: John Wiley & Sons; 2016. pp. 749-772

[13] Arevalos-Sánchez M, Regalado C, Martin SE, Domínguez-Domínguez J, García Almendárez BE. Effect of neutral electrolyzed water and nisin on *Listeria monocytogenes* biofilms and on listeriolys in O activity. Food Control. 2012;**24**(1-2):116-122. DOI: 10.1016/j.foodcont.2011.09.012

[14] Karyotis D, Skandamis PN, Juneja VK. Thermal inactivation of *Listeria monocytogenes* and salmonella spp. in sous-vide processed marinated chicken breast. Food Research International. 2017;**100**(1):894-898. DOI: 10.1016/j.foodres.2017.07.078

[15] Magalhães R, Mena C, Ferreira V, Silva C, Almeida G, Gibbs P, Teixeira P. Bacteria: *Listeria monocytogenes*. Encyclopedia of Food Science. Elsevier; 2014. pp. 450-461

[16] Nyhan M, Begley A, Mutel Y, Qu N, Johnson M, Callanan C. Predicting the combinatorial effects of water activity, pH and organic acids on *Listeria* growth in media and complex food matrices. Food Microbiology. 2018;**74**:75-85. DOI: 10.1016/j.fm.2018.03.002

[17] Phan-Thanh L, Mahouin F, Aligé S. Acid responses of *Listeria monocytogenes*. International Journal of Food Microbiology. 2018;**55**(1–3):121-126

[18] Wiedmann M, Arvik TJ, Hurley RJ, Boor KJ. General stress transcription factor ςB and its role in acid tolerance and virulence of *Listeria monocytogenes*. Journal of Bacteriology. 2018; **180**(14):3650-3656

[19] Miranda RO, Campos-Galvão MEM, Nero LA. Expression of genes associated with stress conditions by *Listeria monocytogenes* in interaction with nisin producer *Lactococcus lactis*. Food Research International. 2018;**105**:897-904. DOI: 10.1016/j.foodres.2017.12.030

[20] Supa-amornkul S, Chantratita W, Srichunrusami C, Janchompoo P, Chaturongakul S. *Listeria monocytogenes* MerR-like regulator NmlRlm: Its transcriptome and role in stress response. Foodborne Pathogens and Disease. 2016;**13**(7):369-378. DOI: 10.1089/fpd.2015.2101

[21] Farber JM, Coates F, Daley E. Minimum water activity requirements for the growth of *Listeria monocytogenes*. Letters in Applied Microbiology. 1992;**15**(3):103-105. DOI: 10.1111/j.1472-765X.1992.tb00737.x

[22] Schvartzman MS, Maffre A, Tenenhaus-Aziza F, Sanaa M, Butler F, Jordan K. Modelling the fate of *Listeria monocytogenes* during manufacture and ripening of smeared cheese made with pasteurised or raw milk. International Journal of Food Microbiology. 2011; **145**:S31-S38. DOI: 10.3389/fcimb.2014.00090

[23] García-Díez J, Patarata L. Influence of salt level, starter culture, fermentable carbohydrates, and temperature on the behaviour of *L. monocytogenes* in sliced chouriço during storage. Acta Alimentaria. 2017;**46**(2):206-213

[24] Bergholz TM, den Bakker HC, Fortes ED, Boor KJ, Wiedmann M. Salt stress phenotypes in *Listeria monocytogenes* vary by genetic lineage and temperature. Foodborne Pathogens and Disease. 2010;**7**:1537-1549. DOI: 10.1089/fpd.2010.0624

[25] Duché O, Trémoulet F, Glaser P, Labadie J. Salt stress proteins induced in *Listeria monocytogenes*. Applied and Environmental Microbiology. 2002;**68**:1491-1498. DOI: 10.1128/AEM.68.4.1491-1498.2002

[26] Gardan R, Duché O, Leroy-Sétrin S, Labadie J. Role of ctc from *Listeria monocytogenes* in osmotolerance. Applied and Environmental Microbiology. 2003;**69**(1):154-161. DOI: 10.1128/AEM.69.1.154-161.2003

[27] Bayles DO, Wilkinson BJ. Osmoprotectants and cryoprotectants for *Listeria monocytogenes*. Letters in Applied Microbiology. 2000;**30**(1):23-27. DOI: 10.1046/j.1472-765x.2000.00646.x

[28] Sleator RD, Wouters J, Gahan CG, Abee T, Hill C. Analysis of the role of OpuC, an osmolyte transport system, in salt tolerance and virulence potential of *Listeria monocytogenes*. Applied and Environmental Microbiology. 2001;**67**:2692-2698. DOI: 10.1128/AEM.67.6.2692-2698.2001

[29] Dykes GA, Moorhead SM. Survival of osmotic and acid stress by *Listeria monocytogenes* strains of clinical or meat origin. International Journal of Food Microbiology. 2000;**56**:161-166. DOI: 10.1016 / S0168-1605 (99) 00205-6

[30] Gormley FJ, Little CL, Grant KA, De Pinna E, McLauchlin J. The microbiological safety of ready-to-eat specialty meats from markets and specialty food shops: A UK wide study with a focus on salmonella and *Listeria monocytogenes*. Food Microbiology. 2010;**27**:243-249. DOI: 10.1016/j.fm.2009.10.009

[31] Tsigarida E, Skandamis P, Nychas GJ. Behaviour of *Listeria monocytogenes* and autochthonous flora on meat stored under aerobic, vacuum and modified atmosphere packaging conditions with or without the presence of oregano essential oil at 5°C. Journal of Applied Microbiology. 2000;**89**(6):901-909. DOI: 10.1046/j.1365-2672.2000.01170.x

[32] Miladi H, Bakhrouf A, Ammar E. Cellular lipid fatty acid profiles of reference and food isolates *Listeria monocytogenes* as a response to refrigeration and freezing stress. Journal of Food Biochemistry. 2013;**37**(2):136-143. DOI: 10.1111/j.1745-4514.2011.00607.x

[33] Borezee E, Pellegrini E, Berche P. OppA of *Listeria monocytogenes*, an oligopeptide-binding protein required for bacterial growth at low temperature and involved in intracellular survival. Infection and Immunity. 2000;**68**(12):7069-7077

[34] Wemekamp-Kamphuis HH, Sleator RD, Wouters JA, Hill C, Abee T. Molecular and physiological analysis of the role of osmolyte transporters BetL, Gbu, and OpuC in growth of *Listeria monocytogenes* at low temperatures. Applied and Environmental Microbiology. 2004;**70**(5):2912-2918. DOI: 10.1128/AEM.70.5.2912-2918.2004

[35] Saraiva C, Fontes MC, Patarata L, Martins C, Cadavez V, Gonzales-Barron U. Modelling the kinetics of *Listeria monocytogenes* in refrigerated fresh beef under different packaging atmospheres. LWT Food Science and Technology. 2016;**66**:664-671. DOI: 10.1016/j.lwt.2015.11.026

[36] Ariyapitipun T, Mustapha A, Clarke AD. Survival of *Listeria monocytogenes* Scott a on vacuum-packaged raw beef treated with polylactic acid, lactic acid, and nisin. Journal of Food Protection. 2000;**63**(1):131-136

[37] Grau FH, Vanderlinde PB. Growth of *Listeria monocytogenes* on vacuum-packaged beef. Journal of Food Protection. 1990;**53**(9):739-741

[38] De Cesare A, Valero A, Lucchi A, Pasquali F, Manfreda G. Modeling growth kinetics of *Listeria monocytogenes* in pork cuts from packaging to fork under different storage practices. Food Control. 2013;**34**:198-207

[39] Gouveia AR, Alves M, de Almeida JMM, Monteiro-Silva F, González-Aguilar G, Silva JA, Saraiva C. The antimicrobial effect of essential oils against *Listeria monocytogenes* in sous vide cook-chill beef during storage. Journal of Food Processing and Preservation. 2017; **41**(4):39-44. DOI: 10.1111/jfpp.13066

[40] Solomakos N, Govaris A, Koidis P, Botsoglou N. The antimicrobial effect of thyme essential oil, nisin, and their combination against *Listeria monocytogenes* in minced beef during refrigerated storage. Food MicrobiologyFood microbiology. 2008;**25**(1):120-127. DOI: 10.1016/j.fm.2007.07.002

[41] El Abed N, Kaabi B, Smaali MI, Chabbouh M, Habibi K, Mejri M, Nejib M, Marzouki MN, Ahmed SBH. Chemical composition, antioxidant and antimicrobial activities of Thymus capitata essential oil with its preservative effect against *Listeria* m*onocytogenes* inoculated in minced beef meat. Journal of Evidence-Based Integrative Medicine. 2014;**204**:29-36

[42] Pesavento G, Calonico C, Bilia AR, Barnabei M, Calesini F, Addona R, et al. Antibacterial activity of oregano, Rosmarinus and thymus essential oils against *Staphylococcus aureus* and *Listeria monocytogenes* in beef meatballs. Food Control. 2015;**54**:188-199. DOI: 10.1016/j.foodcont.2015.01.045

[43] Beverlya RL, Janes ME, Prinyawiwatkula W, No HK. Edible chitosan films on ready-to-eat roast beef for the control of *Listeria monocytogenes*. Food Microbiology. 2008;**25**(3):534-537. DOI: 10.1016/j.fm.2007.11.002

[44] Jiang Z, Neetoo H, Chen H. Efficacy of freezing, frozen storage and edible antimicrobial coatings used in combination for control of *Listeria monocytogenes* on roasted Turkey stored at chiller temperatures. Food MicrobiologyFood microbiology. 2011;**28**(7):1394-1401. DOI: 10.1016/j.fm.2011.06.015

[45] Esteves A, Santos C, Silva JÁ, Patarata L, Martins C. Effect of some chemical treatments on *L. monocytogenes* and *S. aureus* present on meat surface. Veterinária Técnica. 2000;**10**(6):22-25

[46] Gialamas H, Zinoviadou KG, Biliaderis CG, Koutsoumanis KP. Development of a novel bioactive packaging based on the incorporation of lactobacillus sakei into sodium-caseinate films for controlling *Listeria monocytogenes* in foods. Food Research International. 2010;**43**(10):2402-2408. DOI: 10.1016/j.foodres.2010.09.020

[47] McCallin S, AlamSarker S, Barretto C, Sultana S, Berger B, Huq S, et al. Safety analysis of a Russian phage cocktail: From met a genomic analysis to oral application in healthy human subjects. Virology. 2013;**443**:187196. DOI: 10.1016/j.virol.2013.05.022

[48] Bai J, Kim YT, Ryu S, Lee JH. Biocontrol and rapid detection of food-borne pathogens using bacteriophages and endolysins. Frontiers in Microbiology. 2016;**7**:474. DOI: 10.3389/fmicb.2016.00474

[49] Soni KA, Nannapaneni R. Removal of *Listeria monocytogenes* biofilms with bacteriophage P100. Journal of Food Protection. 2010;**73**:1519-1524

[50] Chibeu A, Agius L, Gao A, Sabour PM, Kropinski AM, Balamuru-gan S. Efficacy of bacteriophage LISTEX P100 combined with chemical antimicrobials in reducing *Listeria monocytogenes* in cooked Turkey and roast beef. International Journal of Food Microbiology. 2013;**167**:208-214. DOI: 10.1016/j.ijfoodmicro.2013. 08.018

[51] Carlton RM, Noordman WH, Biswas B, DeMeester ED, Loessner MJ. Bacteriophage P100 for control of *Listeria monocytogenes* in foods: Genome sequence, bioinformatics analyses, oral toxicity study, and application. Regulatory Toxicology and Pharmacology. 2005;**43**: 301-312. DOI: 10.1016/j.yrtph.2005. 08.005

[52] Kim JW, Siletzky RM, Kathariou S. Host ranges of listeria-specific bacteriophages from the Turkey processing plant environment in the United States. Applied and Environmental Microbiology. 2008;**74**:6623-6630. DOI: 10.1128/AEM.01282-08

[53] Bigot B, Lee WJ, Mcintyre L, Wilson T, Hudson JA, Billington C, et al. Control of *Listeria monocytogenes* growth in a ready-to-eat poultry product using a bacteriophage. Food Microbiology. 2011;**28**. DOI: 1448-1452. DOI:10.1016/j.fm.2011.07.001

[54] Guenther S, Huwyler D, Richard S, Loessner MJ. Virulent bacteriophage for efficient biocontrol of *Listeria monocytogenes* in ready-to-eat foods. Applied and Environmental Microbiology. 2009;**75**:93-100. DOI: 10.1128/AEM.01 711-08

[55] Anany H, Chen W, Pelton R, Griffiths MW. Biocontrol of *Listeria monocytogenes* and *Escherichia coli* O157:H7 in meat by using phages immobilized on modified cellulose membranes. Applied and Environmental Microbiology. 2011;**77**:6379-6387. DOI: 10.1128/AEM.05493-11

[56] Ross T, McMeekin TA. Modeling microbial growth within food safety risk assessments. Risk Analysis. 2003;**23**:179-197

[57] Baranyi J, Roberts TA. Mathematics of predictive food microbiology. International Journal of Food Microbiology. 1995;**26**:199-218. DOI: 10.1016/0168-1605(94)00121-L

[58] Huang L. Growth kinetics of *Listeria monocytogenes* in broth and beef frankfurters - determination of *lag* phase duration and exponential growth rate under isothermal conditions. Journal of Food Science. 2008;**73**:E235-E242

[59] Rosso L, Bajard S, Flandrois JP, Lahellec C, Fournaud J, Veit P. Differential growth of *Listeria monocytogenes* at 4 and 8°C: Consequences for the shelf-life of chilled products. Journal of Food Protection. 1996;**59**:944-949

[60] Whiting RC. Microbial modeling in foods. Critical Reviews in Food Science and Nutrition. 1995;**35**(6):467-494

[61] Cornu M, Billoir E, Bergis H, Beaufort A, Zuliani V. Modeling microbial competition in food: Application to the behavior of *Listeria monocytogenes* and lactic acid flora in pork meat products. Food Microbiology. 2011;**28**(4):639-647. DOI: 10.1016/j.fm.2010.08.007

[62] Couvert O, Pinon A, Bergis H, Bourdichon F, Carlin F, Cornu M, et al. Validation of a stochastic modelling approach for *Listeria monocytogenes* growth in refrigerated foods. International Journal of Food Microbiology. 2010;**144**:236-242. DOI: 10.1016/j.ijfoodmicro.2010.09.024

[63] Pouillot R, Lubran MB. Predictive microbiology models vs. modeling microbial growth within *Listeria monocytogenes* risk assessment: What parameters matter and why. Food Microbiology. 2011;**28**(4):720-726. DOI: 10.1016/j.fm.2010.06.002

[64] Pathogen Modeling Program (PMP) available from: https://pmp.errc.ars.usda.gov/PMPOnline.aspx [Accessed: 2018-06-22]

[65] SymPrevius (http://www.symprevius.net/)

[66] ComBase (http://www.combase.cc/index.php/en/) [Accessed: 2018-06-22]

[67] Augustin JC, Bergis H, Midelet-Bourdin G, Cornu M, Couvert O, Denis C, et al. Design of challenge testing experiments to assess the variability of *Listeria monocytogenes* growth in foods. Food Microbiology. 2011;**28**:746-754. DOI: 10.1016/j.fm.2010.05.028

Section 3

Listeria monocytogenes in Medicine Research

Chapter 4

Listeria monocytogenes in Medical Research

Nihed Ben Halima

Additional information is available at the end of the chapter

http://dx.doi.org/10.5772/intechopen.74840

Abstract

Bacteria are known to produce compounds of high value such as secondary metabolites used in biotechnological applications. It is therefore worthwhile to think how to exploit a pathogenic bacterium, e.g., *Listeria monocytogenes*, to be an effective source of bioactive compound used in particular in medicinal purposes. *Listeria monocytogenes* is considered as an acute contaminated bacterium in foods and could be a causal agent of food-borne diseases. This bacterium is the causal agent of listeriosis, a grave disease, caused by eating contaminated food. Although, *L. monocytogenes* is a pathogenic microorganism that threatens the progress of food industry, it would be also a reservoir of secondary metabolites such as antibiotics and other metabolites of economic importance when appropriate strain improvement will be addressed. This section would discuss in brief the negative and positive features of *L. monocytogenes* as either a pathogenic bacterium or an important microorganism in medical research.

Keywords: productive strain, metabolic diseases, food-borne pathogens, valorization, biotechnological applications

1. Introduction

Listeria monocytogenes is a Gram-positive bacterium that is recognized as a facultative intracellular pathogen. An intracellular growth could therefore be observed among food and clinical strains of such bacterium [1].

L. monocytogenes is a well-known food-borne pathogen, which has been found in many fresh and processed foods, and it is widely distributed in nature. In fact, this organism is able to survive in extreme environments including elevated osmolarity, cold, and acid shocks. Understanding the key stress adaptation is important for a better control of *L. monocytogenes* in food as well as in its resulted health diseases. Some authors have reported studies on the protein patterns expressed in response to salt shock in *L. monocytogenes* [2–4].

L. monocytogenes could enter the food chain and lead to severe disseminated infection, as listeriosis, and this feature is very likely due to its ability to survive in both reduced temperature and high-salt conditions [5].

L. monocytogenes can be transmitted to humans through ingestion of contaminated food, particularly ready-to-eat meat, seafood, and dairy products [6].

Listeria monocytogenes is therefore an interesting microorganism to be studied mainly in food and medical research.

The present study focuses on reviewing and describing some important features of *Listeria monocytogenes* from different published reports.

2. Adaptation to salt stress and protein patterns of *L. monocytogenes*

Listeria monocytogenes is able to tolerate salt stress. The mechanism of adaptation to increased salt concentration by this bacterium could be due to intracellular accumulation of compatible solutes. Indeed, the compatible solutes, such as carnitine and glycine betaine, protect the cell from deleterious effects of the external osmolarity and prevent water loss [7, 8].

Comparison of the salt-induced protein patterns of *L. monocytogenes* strain LO28 grown in a rich medium or in a chemically defined medium shows clear differences between these two media as reported by Duché et al. [2]. The later report revealed that the NaCl stress response of *L. monocytogenes* is a complex process. Indeed, there is synthesis of different proteins more or down-expressed in the presence of salt and either directly related or not to the salt stress response of *L. monocytogenes*. The protein pattern analysis revealed the synthesis of three proteins of the general metabolism (AckA, PdhD, and S6), which was modified after salt stress, but does not seem to be directly related to the salt stress response of *L. monocytogenes*. However, two proteins more expressed (GbuA and Ctc) in the presence of salt seem to be directly related with the response to salt stress of *L. monocytogenes*.

The report of Dussurget et al. [9] revealed the presence of an *L. monocytogenes*-specific putative gene encoding a bile salt hydrolase (BSH) and demonstrated that BSH is a novel PrfA-regulated *L. monocytogenes* virulence factor involved in the intestinal and hepatic phases of listeriosis [9].

Protein patterns of *L. monocytogenes* were also analyzed by proteomic analysis in comparison with mode of growth either in biofilm or in planktonic mode [10]. The results showed a significant variation of the protein patterns of *L. monocytogenes* between the two growth conditions. The study indicated in particular that the biofilm development is probably controlled by specific regulation of protein expression involved at various levels of cellular physiology [10].

3. Adaptation of *L. monocytogenes* to reduced temperature

Controlling *L. monocytogenes* in food has become a major preoccupation in the food industry and storage. For this end, many reports were investigated to determine the efficacy of natural drugs such as essential oils to inhibit the growth of such microorganism. In this regard, beef meat including plant essential oil such as that from lemon (*Citrus limon*) is an interesting target during refrigerated storage as contamination of beef meat by food spoilage and food-borne pathogens is considered one of the major problems to the progress of food industry. Indeed, the addition of such essential oil could substantially delay the growth of *L. monocytogenes* ATCC 19117 in raw minced beef meat under storage at 4°C [11].

Listeria monocytogenes is one of the most important psychrotrophic food pathogens, which is responsible for food-borne illnesses in particular listeriosis that has been known as one of the emerging zoonotic diseases nowadays [12, 13]. *L. monocytogenes* could be related to anaerobically packed cooked meat products and shelf-life failures of conserved foods.

Foods are exposed to contamination by several bacteria such as those reported as the causal agents of food-borne diseases, i.e., *Staphylococcus aureus*, *Salmonella*, *Escherichia coli* O157:H7, and *Listeria monocytogenes* [14, 15].

The latter pathogen is the causal agent of listeriosis, a potential grave disease, and is often fatal in susceptible individuals, caused by eating contaminated food in particular with *Listeria monocytogenes* [16].

To prevent food contamination during the production, sale, and distribution and to extend the shelf-life time of raw and/or processed foods, additives should be used. However, the safety aspects of the synthetic additives could be paid attention, as these chemical preservatives are considered toxic and responsible for many carcinogenic and teratogenic attributes [17].

Natural products such as plants and herbs and naturally derived compounds are regarded as new alternatives to prevent the proliferation of pathogens, e.g., strains of *Listeria monocytogenes* in food [11, 18].

A particular interest has been focused on the potential application of plant essential oils such as those from lemon [11] and thyme [18] as safer additives for meat, in particular minced beef meat during refrigerated storage.

4. Some traits highlighted in *L. monocytogenes*

Food and clinical strains of *Listeria monocytogenes* were used in the report of Kanki et al. [1] to study specific alleles in particular the activities of listeriolysin O (LLO) and phospholipases PlcA and PlcB that are known to promote rupture of the phagocytic vacuole, besides initial intracellular bacterial growth in Caco-2 cells.

In fact, the aim of this report [1] is to investigate whether *Listeria monocytogenes* strains differ in their ability to escape from the primary phagosome after internalization into human intestinal epithelial cells.

The results showed differences among LLO and PrfA truncation mutants, but there are no differences in LLO activities between food and clinical strains of *Listeria monocytogenes* or among their serotypes. The same authors [1] concluded that LLO and PrfA mutants of *Listeria monocytogenes* exert a significant effect on intracellular growth, although it was unclear from this study whether PlcA and PlcB alleles affect escape from vacuoles.

Their study put down that low-virulence *L. monocytogenes* strains associated with escape ability from the primary vacuoles are not widely distributed among food strains [1].

Other reports of Camilli et al. [19] showed that the plcA gene of *Listeria monocytogenes* encoding a secreted phosphatidylinositol-specific phospholipase C (PI-PLC) plays important roles in pathogenesis.

L. monocytogenes is responsible for food-borne infections that can cause many health perturbations, e.g., septicemia and meningitis in immunocompromised individuals as well as stillbirth and miscarriage in pregnant women [6].

L. monocytogenes invades intestinal epithelial cells after passing through the stomach, via internalin A (InlA), which interacts with the E-cadherin receptor on the epithelial cell surface [20]. During invasion, *L. monocytogenes* cells become engulfed within a phagocytic vacuole constituting the primary phagosome. Phospholipases PlcA and PlcB, encoded by *plcA* and *plcB*, respectively, and listeriolysin O (LLO), encoded by *hly*, are responsible to promote rupture of the internalized vacuole and bacterial escape into the cytoplasm [21]. Indeed, LLO favors introduction of pores in the membrane of the phagosome, and phospholipases (PlcA and PlcB) help the damage of the membrane, leading the vacuoles to lyse [22].

The replication of *Listeria monocytogenes* occurs within the cytoplasm, and its motility is mediated by ActA that induces actin to polymerize forming actin comet tails around the bacteria. Upon reaching the plasma membrane, *Listeria monocytogenes* invades and internalizes into neighboring cells and disseminates the infection to other cells; such dissemination is called cell-to-cell spread [22]. When invading neighboring cells, *L. monocytogenes* could be localized in a double-membrane (secondary) vacuole, which can be lysed by LLO, PlcA, and PlcB [23]. Camargo et al. [22] denoted that the *Listeria* genes encoding virulence factors for intestinal translocation, namely, *hly*, *plcA*, *plcB*, *actA*, and their regulator *prfA*, reside in the *Listeria* pathogenicity island 1 (LIPI-1).

The virulence gene *inlA* requires activation of the PrfA, which is the virulence regulator for transcription. *L. monocytogenes* expresses the virulence factors and crosses the intestinal barrier to be then carried by the lymph or blood to the mesenteric lymph nodes, spleen, and liver [20].

Many mutant strains of *L. monocytogenes* are widely distributed among food strains especially those of truncated InlA [22]. In fact, mutants of *inlA* with nucleotide substitutions introduce that premature stop codons (PMSCs) express an inactive, truncated InlA.

PMSCs in *inlA* could be presented in isolates of serotypes 1/2a, 1/2b, and 1/2c, but not 4b, and *inlA* PMSCs are most commonly present in food isolates, with few detected in clinical isolates [24]. Orsi et al. [24] indicated that the heterogeneity among the serotypes of clinical isolates of *L. monocytogenes* (e.g., predominance of serotype 4b) may be explained by *inlA* PMSCs.

L. monocytogenes would comprise also hypervirulent and hypovirulent clones [25]. Except for *L. monocytogenes* hypovirulent food isolates with *inlA* PMSCs, attenuated *L. monocytogenes* strains could be isolated from food environments and harbored mutations in the genes encoding PrfA, InlB, PlcA, LLO, and ActA [26, 27].

In addition to being an important pathogen for humans and animals, *L. monocytogenes* is also being regarded and developed as a novel vaccine platform, in particular for tumor immunotherapy [28].

The infection of *L. monocytogenes* has naturally triggered robust CD8+ T-cell responses due essentially to its constitutive intracellular life cycle [29].

L. monocytogenes could be promising bacteria for immunotherapy platform, which could be illustrated by the >15 active or completed clinical trials when using attenuated *L. monocytogenes* for the treatment of a variety of cancers (http://clinicaltrials.gov).

The exact mechanisms by which *L. monocytogenes* triggers cell-mediated immunity remain unclear, and rigorous efforts are needed to further exploit *L. monocytogenes* as an effective agent for cancer therapy.

During infection, *L. monocytogenes* targets antigen-presenting cells, and it delivers antigens directly to the class I major histocompatibility complex (MHC) presentation pathway, due to its cytosolic localization. This bacterium is genetically tractable, facilitating pathogen attenuation for clinical safety as well as the ability to engineer the pathogen to express tumor antigens of interest [28].

There are two different *L. monocytogenes*-based immunotherapeutic platforms from Aduro BioTech [30] and Advaxis [31] indicating that cytosolic access is necessary for triggering cell-mediated immunity, while cell-to-cell spread of the pathogen is not, thereby ensuring vaccine safety [30, 31].

L. monocytogenes infection is a complex process that could affect the death pathways of different host cells, including programmed and non-programmed cell death.

The review of McDougal and Sauer [32] highlighted the mechanisms of modulation of cell death and its implications on *L. monocytogenes* acute infection, as well as the generation of adaptive immunity. Indeed, in this review [32], the influences of host cell death pathways were discussed, including necrosis and necroptosis, apoptosis, and inflammasome-mediated pyroptosis on both *L. monocytogenes* virulence and *L. monocytogenes*-induced immunity.

Thus, it is important to understand how cell death modulation during *L. monocytogenes* infection could lead to novel insight into therapeutic approaches for the treatment of infection and the development of vaccine strains as cancer immunotherapies. The role of *L. monocytogenes* as a tumor immunotherapy platform would open other questions on how critical cell death pathways are that influence the priming and quality of cell-mediated immune responses [32].

Other reports from Buyck et al. [33] described the presentation, diagnosis, and treatment of *Listeria monocytogenes* sepsis in an older patient and presented a short literature review about listeriosis and the importance of safe food practices [33].

5. Current and future developments

The starting point for this chapter is the characterization of *Listeria monocytogenes* as a virulent organism and as a possible turn of such pathogen to be effective as preventive or curative agent such as vaccine.

As any other organisms/microorganisms, *Listeria monocytogenes* could follow a strategy to have an improved strain of such bacterium. For this end, screening for productive strains and strain improvement in biotechnological organisms of *Listeria monocytogenes* would be the focus of this part of the current chapter. *Listeria monocytogenes* could be beneficial bacteria in modern industrial microbiology and biotechnology, and the strategy used to achieve such purpose would be summarized below.

On the one hand, it is important to scratch for sources of microorganisms, e.g., strains of *Listeria monocytogenes* used in biotechnology. To achieve such goal, two important steps to be borne in mind are (a) literature search and culture collection supply and (b) isolation de novo of organisms producing metabolites of economic importance. Some general screening methods are described as three important points: (i) enrichment with the substrate utilized by the organism being sought, (ii) enrichment with toxic analogues of the substrate utilized by the organism being sought, and (iii) testing microbial metabolites for bioactive activity. The later point would involve such as:

- Testing for antimicrobial activity
- Testing for enzyme inhibition
- Testing for morphological changes in fungal test organisms
- Conducting animal tests on the microbial metabolites

On the other hand, it is now important to focus on strain improvement when the above steps were clearly understood and investigated.

It is also important to open a parenthesis talking about the fact that:

The ability of any organism, e.g., *Listeria monocytogenes*, to make any particular product, e.g., bioactive compounds, is based on its capability for the secretion of a particular set of enzymes. Enzymes are the key factors of organisms' life. Moreover, the production of the enzymes depends ultimately on the genetic makeup of the organisms. How to improve organism strains? The answer of this question could be put down with five condition procedures as follows:

1. Regulating the enzymes' activities secreted by the organism
2. Increasing the permeability of the organism in the case of searching metabolites secreted extracellularly, so that the microbial products can find this way more easily outside the cell
3. Selecting strains from natural variants
4. Modifying the existing genetic apparatus in a producing organism
5. Using recombinant DNA technology or genetic engineering

The two latter possible procedures, namely, modification of the existing genetic apparatus without the introduction of foreign DNA and introducing new genetic properties into the organism being sought by using foreign DNA, will be discussed below in general for any industrial organism.

The manipulation of the genome of industrial organisms for the purpose of strain improvement may be done either by manipulation not involving foreign DNA or that involving foreign DNA.

On the one hand, the general procedure of the genome manipulations not involving foreign DNA or bases could be predicated on conventional mutation. Indeed, the nature of conventional mutation can be originated from physical agents with ionizing radiations and/or ultraviolet light, besides chemical mutagens.

These chemical mutagens, for instance, may be divided into three groups:

i. Chemical mutagens that act on DNA of resting or nondividing organisms (chemicals acting on resting DNA) such as nitrous acid, alkylating agents, NTG (nitrosoguanidine), and nitrogen mustards
ii. DNA analogues which may be incorporated into DNA during replication (base analogues)
iii. Chemical mutagens that cause frameshift mutations (also known as intercalating agents)

Whatever mutagens are used (physical or chemical agents), the choice of either mutagen should satisfy the final effective purpose, and it is also important to bear in mind the practical isolation of mutants. For this end, some general related practices should be taken appropriately such as exposing organisms to the mutagen, selection for mutants, and screening.

Moreover, the isolation of auxotrophic mutants can be of great importance in modern industrial microbiology and biotechnology as these mutants are frequently used in industries for the production, for instance, of essential amino acids. We can note that in contrast to the wild-type or prototrophic organisms that possess all the enzymes needed to synthesize all growth requirements, auxotrophic mutants are those which lack the enzymes to manufacture certain required nutrients; consequently, such nutrients must therefore be added to the growth medium.

On the other hand, the general procedure of strain improvement methods involving foreign DNA or bases could be predicated on many genetic-changing methods such as transduction, transformation, conjugation, parasexual recombination, protoplast fusion, site-directed mutation, and metabolic engineering.

6. Conclusion

In summary, this chapter highlighted primarily that *Listeria monocytogenes* is a well-known pathogenic bacterium particularly to humans and animals causing severe infections especially those from food-borne such as listeriosis. In fact, *L. monocytogenes* could enter the food chain and lead to food-borne illness even at refrigerated temperatures. The pervasiveness of this food spoilage microorganism is due, in part, to its ability to tolerate environments including reduced temperatures, elevated osmolarity, and acid shocks. Consequently, an adequate surveillance system for safe food practices, handling, and storage needs to be established to control *L. monocytogenes* and, thus, should be taken into consideration for healthier world.

Although *L. monocytogenes* is a pathogenic microorganism that threatens the progress of food industry, it would be also a reservoir of secondary metabolites such as antibiotics and other metabolites of economic importance when appropriate strain improvement is addressed. Furthermore, understanding the mechanism of action of *L. monocytogenes* with regard to the infection as well as the immunity could provide critical insights into novel therapeutics for the treatment of infection, as well as the development of vaccine strains as tumor immunotherapies. Thus, it could also refine our use of pathogenic microbes such as *L. monocytogenes* as beneficial microorganisms used in vaccines and cancer immunotherapy.

Author details

Nihed Ben Halima

Address all correspondence to: nihedbenhalima@gmail.com

Faculty of Medicine of Sfax, University of Sfax, Sfax, Tunisia

References

[1] Kanki M, Naruse H, Kawatsu K. Comparison of listeriolysin O and phospholipases PlcA and PlcB activities, and initial intracellular growth capability among food and clinical strains of *Listeria monocytogenes*. Journal of Applied Microbiology. 2018;**124**:899-909

[2] Duché O, Trémoulet F, Namane A, The European Listeria Genome Consortium, Labadie J. A proteomic analysis of the salt stress response of Listeria monocytogenes. FEMS Microbiology Letters. 2002;**215**:183-188

[3] Duché O, Tremoulet F, Glaser P, Labadie J. Salt stress proteins induced in *Listeria monocytogenes*. Applied and Environmental Microbiology. 2002;**68**:1491-1498

[4] Esvan H, Minet J, Laclie C, Cormier M. Protein variations in *Listeria monocytogenes* exposed to high salinities. International Journal of Food Microbiology. 2000;**55**:151-155

[5] Ferreira V, Wiedmann M, Teixeira P, Stasiewicz MJ. *Listeria monocytogenes* persistence in food-associated environments: Epidemiology, strain characteristics, and implications for public health. Journal of Food Protection. 2014;**77**:150-170

[6] Lianou A, Sofos JN. A review of the incidence and transmission of *Listeria monocytogenes* in ready-to-eat products in retail and food service environments. Journal of Food Protection. 2007;**70**:2172-2198

[7] Gerhardt PN, Tombras Smith L, Smith GM. Osmotic and chill activation of glycine betaine porter II in *Listeria monocytogenes* membrane vesicles. Journal of Bacteriology. 2000; **182**:2544-2550

[8] Sleator RD, Wouters J, Gahan CG, Abee T, Hill C. Analysis of the role of OpuC, an osmolyte transport system, in salt tolerance and virulence potential of *Listeria monocytogenes*. Applied and Environmental Microbiology. 2001;**67**:2692-2698

[9] Dussurget O, Cabanes D, Dehoux P, Lecuit M, The European Listeria Genome Consortium, Buchrieser C, Glaser P, Cossart P. *Listeria monocytogenes* bile salt hydrolase is a PrfA-regulated virulence factor involved in the intestinal and hepatic phases of listeriosis. Molecular Microbiology. 2002;**45**:1095-1106

[10] Trémoulet F, Duche O, Namane A, Martinie B, Consortium TELG, Labadie JC. Comparison of protein patterns of *Listeria monocytogenes* grown in biofilm or in planktonic mode by proteomic analysis. FEMS Microbiology Letters. 2002;**210**:25-31

[11] Ben Hsouna A, Ben Halima N, Smaoui S, Hamdi N. *Citrus lemon* essential oil: Chemical composition, antioxidant and antimicrobial activities with its preservative effect against *Listeria monocytogenes* inoculated in minced beef meat. Lipids in Health and Disease. 2017;**16**:146

[12] Alam MS, Costales M, Cavanaugh C, Pereira M, Gaines D, Williams K. Oral exposure to *Listeria monocytogenes* in aged IL-17RKO mice: A possible murine model to study listeriosis in susceptible populations. Microbial Pathogenesis. 2016;**99**:236-246

[13] Castro SM, Kolomeytseva M, Casquete R, Silva J, Queirós R, Saraiva JA, Teixeira P. Biopreservation strategies in combination with mild high pressure treatments in traditional Portuguese ready-to-eat meat sausage. Food Bioscience. 2017;**19**:65-72

[14] Ben Hsouna A, Trigui M, Ben Mansour R, Jarraya RM, Damak M, Jaoua S. Chemical composition, cytotoxicity effect and antimicrobial activity of *Ceratonia siliqua* essential oil with preservative effects against Listeria inoculated in minced beef meat. International Journal of Food Microbiology. 2011;**148**:66-72

[15] Rahman A, Kang SC. *In vitro* control of food-borne and food spoilage bacteria by essential oil and ethanol extracts of *Lonicera japonica* Thunb. Food Chemistry. 2009;**116**:670-675

[16] Cornu M, Beaufort A, Rudelle S, Laloux L, Bergis H, Miconnet N, et al. Effect of temperature, water-phase salt and phenolic contents on *Listeria monocytogenes* growth rates on cold-smoked salmon and evaluation of secondary models. International Journal of Food Microbiology. 2006;**106**:159-168

[17] Skandamis P, Koutsoumanis K, Fasseas K, Nychas GJE. Inhibition of oregano essential oil and EDTA on *E. coli* O157:H7. Italian Journal of Food Science. 2001;**13**:55-65

[18] Solomakos N, Govaris A, Koidis P, Botsoglou N. The antimicrobial effect of thyme essential oil, nisin, and their combination against *Listeria monocytogenes* in minced beef during refrigerated storage. Food Microbiology. 2008;**25**:120-127

[19] Camilli A, Tilney LG, Portnoy DA. Dual roles of plcA in *Listeria monocytogenes* pathogenesis. Molecular Microbiology. 1993;**8**:143-157

[20] Vázquez-Boland JA, Kuhn M, Berche P, Chakraborty T, Domínguez-Bernal G, Goebel W, González-Zorn B, Wehland J, Kreft J. *Listeria* pathogenesis and molecular virulence determinants. Clinical Microbiology Reviews. 2001;**14**:584-640

[21] Pizarro-Cerdá J, Kühbacher A, Cossart P. Entry of *Listeria monocytogenes* in mammalian epithelial cells: An updated view. Cold Spring Harbor Perspectives in Medicine. 2012;**2**:a010009

[22] Camargo AC, Woodward JJ, Nero LA. The continuous challenge of characterizing the foodborne pathogen *Listeria monocytogenes*. Foodborne Pathogens and Disease. 2016;**13**:405-416

[23] Travier L, Lecuit M. *Listeria monocytogenes* ActA: A new function for a 'classic' virulence factor. Current Opinion in Microbiology. 2014;**17**:53-60

[24] Orsi RH, den Bakker HC, Wiedmann M. *Listeria monocytogenes* lineages: Genomics, evolution, ecology, and phenotypic characteristics. International Journal of Medical Microbiology. 2011;**301**:79-96

[25] Maury MM, Tsai YH, Charlier C, Touchon M, Chenal-Francisque V, Leclercq A, Criscuolo A, Gaultier C, Roussel S, Brisabois A, Disson O, Rocha EP, Brisse S, Lecuit M. Uncovering *Listeria monocytogenes* hypervirulence by harnessing its biodiversity. Nature Genetics. 2016;**48**:308-313

[26] Roberts A, Chan Y, Wiedmann M. Definition of genetically distinct attenuation mechanisms in naturally virulence-attenuated *Listeria monocytogenes* by comparative cell culture and molecular characterization. Applied and Environmental Microbiology. 2005;**71**:3900-3910

[27] Roche SM, Grépinet O, Kerouanton A, Ragon M, Leclercq A, Témoin S, Schaeffer B, Skoski G, Mereghetti L, Le Monnier A, Velge P. Polyphasic characterization and genetic relatedness of low-virulent *Listeria monocytogenes* isolates. BMC Microbiology. 2012; **12**:304

[28] Le DT, Dubenksy TW, Brockstedt DG. Clinical development of *Listeria monocytogenes*-based immunotherapies. Seminars in Oncology. 2012;**39**:311-322

[29] Bahjat KS, Liu W, Lemmens EE, Schoenberger SP, Portnoy DA, Dubensky TW Jr, Brockstedt DG. Cytosolic entry controls CD8+-T-cell potency during bacterial infection. Infection and Immunity. 2006;**74**:6387-6397

[30] Aduro BioTech. LADD Engineering Listeria Mononcytogenes Bacteria. 2017. Available from http://www.aduro.com/technology/ladd/ [Accessed: May 17, 2017]

[31] Lm Technology—Advaxis. 2017. Available from: https://www.advaxis.com/lm-technology/ [Accessed: Nov 15, 2017]

[32] McDougal CE, Sauer JD. *Listeria monocytogenes*: The Impact of Cell Death on Infection and Immunity. Pathogens. 2018;**7**:8. DOI: 10.3390/pathogens7010008

[33] Buyck G, Devriendt V, Van den Abeele AM, Bachmann C. *Listeria monocytogenes* sepsis in the nursing home community: A case report and short review of the literature. Acta Clinica Belgica. 2018;**Jan 9**:1-5

Chapter 5

Listeria monocytogenes: Potent Clinical Hazard

Prasann Kumar and Shweta Pathak

Additional information is available at the end of the chapter

http://dx.doi.org/10.5772/intechopen.76389

Abstract

Listeria monocytogenes is still the point to be broken by the scientists. In 1967, scientists Gray and Killinger demonstrated, about the presence of *Listeria monocytogenes* and Listeriosis in humans and cattle. *Listeria monocytogenes* was first described by Murray et al., who named it Bacterium Monocytogenes because of a characteristic monocytosis found in infected laboratory rabbits and guinea pigs. In 1927, it was renamed *Listerella hepatolytica* by Pirie who gave its present name in 1940. The first confirmed isolations of the bacterium from infected individuals, following its initial description, were made in 1929 by Gill from sheep and by Nyfeldt from humans. Since then, sporadic cases of listeriosis, have been reported, often in workers in contact with diseased animals. The invasion of peripheral nerve cells and rapid entry into the brain is postulated as a unique characteristic of its virulence.

Keywords: agriculture, biotic, cattle, disease, ecosystem, listeria

1. Introduction

It was the most prominent food-borne and food-causative agent in South America and in some European territories. *Listeria monocytogene* is a Gram-positive type of the facultative intracellular pathogen, which is capable of surviving in presence or absence of oxygen. It is ubiquitous in nature and grows at a minimum temperature of 0°C (typical refrigeration temperatures, greatly increasing its ability to evade control in human foodstuffs) to 50°C, pH, high concentrations of salt or bile, oxidative stress, carbon starvation, and other adverse conditions. Albeit readymade foods, defilement type of raw food, like vegetables, milk, meat, and seafood, is the most abundant contamination zone for such pathogens, not only are these eventually, readymade foods e paradigms of *L. monocytogenes*, they are also the root of the genesis of life-threatening food-borne disease listeriosis, in humans, including young, old,

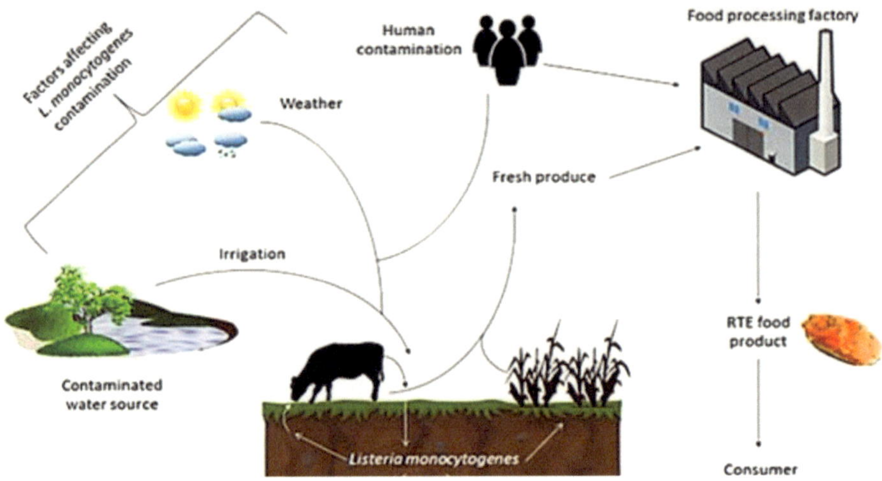

Figure 1. Cross talk of factors responsible for bacteria transmission.

pregnant, and immune-compromised people. Various studies suggested that up to 10% of human gastrointestinal tracts might be colonized by *Listeria monocytogenes*. Approximately 20–30% of the kinds of food-borne listeriosis infections are high risk or may be fatal. It is the highest value observed among all food-borne pathogens overall. The genus *Listeria* comprises nine species: *Listeria monocytogenes, Listeria ivanovii, Listeria innocua, Listeria welshimeri, Listeria seeligeri, Listeria grayi, Listeria marthii* [1], and *Listeria rocourtiae* [2] among which only *L. monocytogenes* is pathogenic to humans and L. ivanovii to animals causing listeriosis (**Figure 1**).

2. Listeriosis

Listeriosis is a disease, irresistibly fundamental spread by the microbes *Listeria monocytogenes*. The sporadic type of listeriosis is found regularly in human beings. The pollute sustenances might be the essential wellspring of tainting. Individuals are inclined to eating defiled, ready-made, or crude sustenances with the microorganisms. It might debilitate with different distinctive locales on a body part, including the cerebrum, spinal rope layers, or the circulation system. Listeriosis is very pervasive, so anybody can be tainted by this sickness, yet the individuals who have debilitated insusceptible framework (for instance, individuals with growth, HIV/Helps, or a transplant), individuals with perpetual liver or kidney ailment, diabetes, or liquor addiction, the neonates, and pregnant ladies are vulnerable to it. In spite of the fact that, dominant part of disease is amicable to who has rose by suppressive Lymphocyte intervened insusceptibility. In any case, the most astounding projection of listerial contamination is generally found in neonates, trailed by people aged 60 years and above (285, 365). In certainty, in a current survey of listeriosis, it was demonstrated that 31 and 22% of the aggregate cases occurred in patients who were aged 60 years and more youthful than 1 month, separately (343). It is evidently get out that of 782 instances of listeriosis in 20 nations 43% were maternal and neonatal diseases, 29% were septicemic contaminations, 24% were focal sensory system contaminations, and 4% were atypical forms. Sensitivity of focal sensory system with *L. monocytogenes* ordinarily showing of meningitis or encephalitis and for the most part gives beginning side effects, including migraine, retching, fever, and discomfort before the presence of central indications of focal sensory system infection. Although 14 cases have been evaluated, in which *L. monocytogenes* assume a significant part to cause cerebrum abscesses in inclined people, particularly in leukemia patients or in renal transplant beneficiaries (94). Meningitis straightforwardly connected with a high death rate (315) in neonates to more seasoned. Symptoms of listeriosis as indicated by reports, listeriosis can influence different body parts, so the manifestations fluctuate from direct to endless. Manifestations are related to prodromal fever and looseness of the bowels alongside other foodborne germs; however, this sort of listeria contamination is not analyzed frequently. Side effects fluctuate with more disease. Pregnant ladies commonly encounter just fever and other influenza-like indications, for example, exhaustion and muscle pain. However, contaminations amid pregnancy can prompt unsuccessful labor, stillbirth, unexpected labor, or hazardous disease of the infant. But pregnant ladies additionally endure manifestations of cerebral pain, firm neck, disarray, loss of adjustment, and shakings alongside fever and muscle pain [3–7].

3. Pragmatic view: resource, occurrence and effect of factors

Our biosphere is a circular pathway for various paradigms of the food chain in which detritivores have their specific importance. The disease-causing pathogens are also included in this group of classification. Their circulation within the biosphere is a major health issue. An agricultural ecosystem is responsible for the increase in transmission of pathogens to the food chain via production of contaminated raw products. Soil is the edaphic factor, which accounts for circulation of *L. monocytogenes*. An important research performed by Welshimer put forward the first evidence that soil is the primary environment for *L. monocytogenes*, and the occurrence of the bacterium was observed in a third of the 12 sampled farms [8]. This report was further reinforced by a team of headed by a scientist named Weis and Seeliger. They contemplated the existence of *L. monocytogenes* in 746 soil samples collected in Southern Germany in which approximately 160 strains of the pathogen were isolated and account for 21.4% incidents of *L. monocytogenes* [9]. The highest incidence was recorded for uncultivated fields and meadows up to 30.8%, while the occurrence was reported to be quite less in cultivated field [10] (**Figure 2**).

The appropriation of the listeria types of spatial variety in urban region soil was accounted around 30% instead of indigenous habitat in US it was 19% [11]. Globally the identification of *L. monocytogenes* has been reported in the wake of examining at a similar site [9, 11]. Besides, the omnipresence of *L. monocytogenes* saw fundamental factor as indicated by the season and classification of condition. Rate of event was additionally amid the period of summer in common habitats while least during this time in urban situations. A nearby to inquire about revealed after their 3 years of study that in vegetable and product cultivates the commonness has been quite recently hostile, it was most noteworthy in winter season [12]. There have been recordings of *L. monocytogenes* in soil tests gathered from little ruminant and cow ranches [21]. The

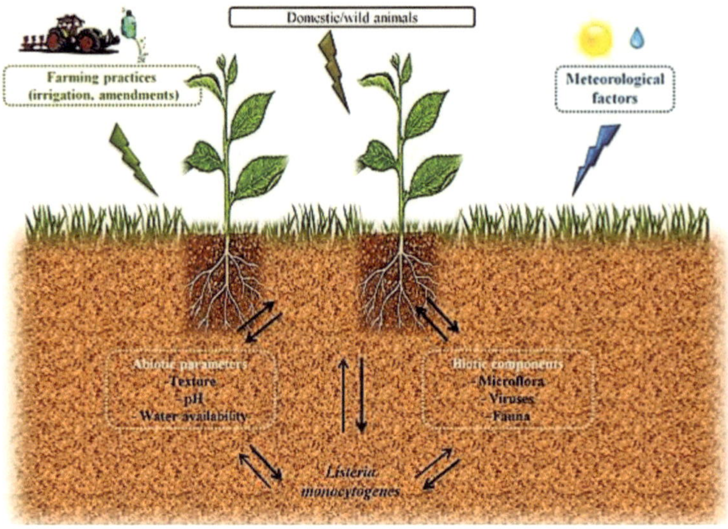

Figure 2. Pragmatic view with special concern to source and occurrence.

recordings show that the number of inhabitants in *L. monocytogenes* in soil is by and large low [13]. This was confirmed in a French across the country review where a PCR examine was performed on a gathering of 1232 soil DNA for the specific identification of *L. monocytogenes* [14]. All examples were underneath the location furthest reaches of 104 g^{-1} of dry soil. Strikingly, a correlation of development construct and sub-atomic location with respect to a subset of 53 crisp soil tests demonstrated that the rate of refined *L. monocytogenes* was 17%, yet just one example showed the discovery of atoms of *L. monocytogenes* and its population was quantified at 2.88104 g^{-1} of dry soil [14]. These overviews obviously exhibit that the dirt is a natural specialty of *L. monocytogenes*, but its population is for the most part low. Worldly and spatial varieties of its event call attention to that ecological factor drive the destiny of *L. monocytogenes* in soil. Soil is a gigantic composite heterogeneous condition made up of natural issue, minerals, frilly plant roots, gases, fluids, and a complex biota, including microorganisms, infections, mesofauna, and macrofauna that together help life. Complex nourishment networks are the characteristic personality of soil that are influenced by subterranean or more ground forms [15]. Because of such properties of soil, it is a significant commonplace errand to disentangle the natural factors that impact the presence of *L. monocytogenes* in soil. For the sake of results from the perception one essential thing turn out that nature of the dirt is a vital factor for event of *L. monocytogenes*. Notwithstanding, in these reports, signs of the dirt attributes are shaggy which is the reason for the difficulty in separating significant data with respect to the connection between soil's qualities and the nearness of *L. monocytogenes*. Examination of listerial population have proven that water is the most fundamental factor for survival of this pathogen [8]. Research to date has revealed that moisture, type of soil, season, the presence of decaying plant material, and possibly the presence of plant root systems, all may influence the growth and survival of *L. monocytogenes* in soil. The soil types examined for their relative ability to support *L. monocytogenes* were distinguished by particle size and organic content. They included a clay loam soil, a sandy loam soil, and a sandy soil. The fertilizer liquid hog manure, solid chicken manure, and inorganic commercial nitrogen-phosphorus-potassium (NPK) were evaluated for any protective effect on the *L. monocytogenes* population. Glucose and peptone solutions were used as nutrient supplements to reveal if *L. monocytogenes* levels in soil were limited by carbon or nitrogen sources. Finally, high and low initial inoculation levels and the presence of normal and reduced (via autoclaving) levels of initial competition were examined for their effect on the establishment of *L. monocytogenes* population. The ultimate purpose of experiments of this nature was to determine what, if any, environmental conditions or agricultural practices might lead to an increased risk of *L. monocytogenes* contamination of food crops. If certain practices or conditions entail a greatly increased risk of elevating *L. monocytogenes* levels in soils, appropriate measures may be taken to offset the risk of crop contamination. The sort of soil influences population flow. Sandy soil speaks to a domain less positive for *L. monocytogenes* than sandy topsoil and earth soils and reduction in the survival in dirt soil than fruitless garden soil was additionally reported [8]. A disadvantage of these investigations is that just a single soil of each kind was utilized. A factual way to deal with the examination of *L. monocytogenes* survival in soils was performed to dodge this constraint of the distributed information [16]. Information for the survival of *L. monocytogenes* in 100 soil microcosms was dissected in light of a far-reaching and point-by-point portrayal of the dirt samples [16]. These results confirmed the survival rates in these kinds of soil. Survival up to 84 days was seen in 71% of the dirt samples tried, while survival did not surpass 14 days in whatever remained of the dirt microcosms. The long-haul survival could be identified with the dirt surface and particularly earth content. The report by [16] and others prove that the pH of the dirt is a noteworthy driver in the spread of *L. monocytogenes* in soil [9, 16].

4. Farming practices and transfer into the soil

Microflora inside the dirt can exceptionally influence the survival of *L. monocytogenes* (**Figure 1**). Connections between *L. monocytogenes* and diverse sorts of protozoa have already been exhibited [17, 18]. Disinfection of soil can prompt an increase in the development of *L. monocytogenes* proposing that the miniaturized scale verdure of the dirt, for example, bacteriophage or protozoa, affect the steadiness of the bacterium, in spite of the fact that this impact has not yet been completely clarified. Reports likewise confirmed that the smaller-scale biota of the dirt assumes a critical part in survival. In their investigation, they, in part, reconstituted sterile soil with social high-impact segments of the dirt small-scale biota and watched that this prompted an abatement in survival at a later time during the test. This proves the likelihood that this abatement might be because of the rivalry between various small-scale greenery that have supplements inside the dirt. Different components, which may influence the survival of the life form in soil, incorporate compound properties and in addition geological and meteorological impacts. Transient components (water system and precipitation) prompted the pollution of pre-gathered spinach. There was a more noteworthy possibility of confining *L. monocytogenes* in the water system than precipitation, and this possibility was most noteworthy within 24 h of this occurrence. Different investigations have confirmed likewise that the water system is a hazard factor for tainting of pre-harvestable nourishments. This is regularly because of the tainting of the water source utilized for the water system in the fields. Alongside water system, the utilization of compost as a manure can build disengagement of *L. monocytogenes* from deliver creation destinations [19, 20]. This is not astounding as several animals are known to supply the bacterium [20]. The Edaphic manifestation of *L. monocytogenes* and other sustenance-borne pathogens hail the well-being concerns ranch and brushing it point toward the dirt is the colossal supply of *L. monocytogenes* and it might be a vector of pathogens to developed plants and cultivated creatures. Therefore, cattle and little ruminants are responsible for the spread of *L. monocytogenes* [21]. Diverse support proposes the possibility that cultivating hones have coordinate concern and effect on the dissemination and implantation of *L. monocytogenes*. During the cultivating and harvest season, tainting revolution surpasses the number of inhabitants in *L. monocytogenes*, which are nourished by dairy cattle raising the issue of sepsis in cows which are listeriosis carriers [20]. Using the same natural squanders or composts without sanitation leads to the transmission of *L. monocytogenes* in soil. The disposed sewage sludge has a low measure of *L. monocytogenes* population (1–240 microbes MPN g^{-1} dry issue), yet malicious utilization of such sewage slime as a compost have encouraged its transmission and implantation into the soil [19, 22]. Per hectare, spreading of one to two tons of muck would bring about 106–108 *L. monocytogenes* every year. There are also reports of the presence of *L. monocytogenes* in the defecated samples of homestead creatures [20, 21]. Additionally, investigation confirmed the nearness and the recurrence in 52 ox-like, goat, and sheep cultivates; the recurrence or presence of *L. monocytogenes* changed from 22–33% in cow-like ranches and from 3 to 18% in goat and sheep ranches [21]. Survival recurrence capacity in fecal waste is wholly restricted to half a month, yet every daily contribution to the storerooms of the barnyard could keep up a steady heap of *L. monocytogenes* [23–25]. *L. monocytogenes* could transmit by means of spreading of untreated waste or fecal ashore and that can sully soil and by refined of vegetables or natural products it move into them and they perform like a supply of *L. monocytogenes*. Such unhygienic septic sullied crude nourishment or instant sustenance of things would endanger the well-being of animals. Amid a similar period, an examination detailed that greatest survival period shifted from 4 days to more than 32 days after land application and such survival of this pathogen relies on this sort of waste. The population of the pathogens influenced or declined by confining the use of waste amid cultivating and development beside the seashore [25]. In vitro condition the survival of this pathogen in cow-like fertilizer changed soil fluctuated from 21 to 43 days and relied upon the measurements and on the temperature of brooding [26]. A factual thinks about information give an intriguing perspective that the natural parameters influence the likelihood of the presence of the pathogen, by the assistance of the information of overview it is likewise certain that a solid relationship between climate, soil properties and the likelihood to recognize Listeria sp. in soil. In a current parallel investigation of the event of sustenance-borne pathogens in five food products grown from the ground homesteads, scene and meteorological elements were related to the recurrence of positive examples [12]. Exhaustively, the occurrence of *L. monocytogenes* was distinguished around 15%, while the recurrence in the soil samples was 9%. This result demonstrated that ecological components like temperature, separation from surface water, streets/urban advancement and field/feed grass, in addition to soil-related parameters (accessible water stockpiling, natural soil) were ecological and topographic variables of significance in locating *L. monocytogenes* [12].

http://dx.doi.org/10.5772/intechopen.76389

Curiously, in this review, the recurrence of identification of *L. monocytogenes* was higher in water tests. The nature of water utilized in the water systems was related to the transmission of pathogens through low-quality water systems [27–29]. Dairy ranches are an example of ecological elements that influence identification of *L. monocytogenes* in watersheds. Wastewater-treated effluents contain *L. monocytogenes* [19, 30–32] with loads fluctuating from 3 to 15 CFU.ml^{-1} [22] to more than 10 3 CFU.ml^{-1}.

5. Biodiversity and prevalence in soil

Biodiversity, short for natural assorted variety, alludes to the majority of the population, species, and groups in a characterized region. As opposed to the more particular term species assorted variety, the term biodiversity was instituted to accentuate the numerous intricate sorts of varieties that exist inside and among creatures at various levels of association. The term biodiversity alludes to the totality of qualities, species, and biological system of the locale. Biodiversity incorporates three progressive levels: hereditary, species, and biological system. Species are unmistakable units of assorted variety each assuming an imperative part in the biological community; it alludes to assortment of species inside a district. The most straightforward measure is more noteworthy the species extravagance more prominent the assorted variety. Biological community incorporates every animal type in addition to all the abiotic factors normal for a district. Biological community decent variety depicts the quantity of specialties, trophic level, and different environmental procedures that support vitality stream, sustenance networks, and the reusing of supplements. Living beings that make up the biotic part of a biological community are generally named autotrophs and heterotrophs, in view of how they get their nourishment or natural supplements in order to survive. *L. monocytogenes* are a facultative heterothophs that use wide variety and range of environment and such virtues represent its specificity. Species traditionally have been described and identified on the basis of morphological criteria. The forms and behavior of organisms adapt to some extent to the resource they exploit and habitats they occupy. Although the classification and identification of bacterial species is not a very easy task as a soil microflora, the term ecotype describes a genetically different population within a species which is adapted to specific environmental conditions. The different ecotypes of a microorganism species may differ in their edaphic, biotic, or microclimatic requirements. It aims at systematic synthesizing of ecology and evolution of microorganisms [33, 34]. Ecotypes are defined as "the smallest groups that (i) show a history of coexistence as separate, ecologically distinctlineages, as inferred from community ecology and (ii) show a prognosis for future coexistence, as inferred from the ecological distinctness of the groups in nature [33]." The phylogenetic structure of *L. monocytogenes* is mind-boggling. Disconnects are gathered in four heredities and major clonal edifices are perceived [35]. These major clonal buildings are dispersed overall [36]. Curiously, there is confirmation that the appropriation of clones and serotypes vary among clinical, sustenance, and natural detaches [36–38]. However, a constraint of these examinations is that the accumulation of separates that were broken down do not completely illustrate the mind-boggling biology of individuals from the species *L. monocytogenes*. Ribotyping of ranch-creature samples of *L. monocytogenes* identified a few subtypes [21, 39]. Confirmation likewise indicated higher identification of specific subtypes in specific test destinations. These outcomes confirm the presence of ecotypes. Notwithstanding, others write about the across-the-board dispersion of PFGE composes paying little heed to their origin [40]. As proposed by Cohan, [33] defining such ecotypes will require a reasonable exhibition that the nature of disengages of the different grouping bunches is really particular.

6. Food processing environments

L. monocytogenes has the dynamic capacity to make due in invert state of the earth, crude nourishment or instant sustenance stuffs are stores of this pathogen and the likelihood of its event is exceptionally sufficiently solid to develop and make due finished a drawn out stretch of time yet how pathogen survive and what is the unmistakable pathway is as yet the striking inquiry. The wonder of constancy is very appropriate for such pathogens, and it is characterized as a specific subtype re-secluded from a similar situation over a broadened time frame. Be that as it may, *L. monocytogenes* have diverse strains and all have their own particular constancy yet, the perserverance of which strain is happening or more grounded in various natural living space is hard to discover why not process condition or others as well. Level headed discussion and difference, dependably the root to opportunity to find out about this pathogenic properties that whether a hereditary variety relate with presence or whether *L. monocytogenes* can colonize under particular great specialties inside a handling domain to stay continuing on. Bit of the work called attention to practically identical reports of research that made the observation that phenotypic attributes with various strains could bolster the steadiness contrasted with non-relentless strains. Consequently, to classify the non-persistent strain alongside persistent [41–43] strain is extremely commonplace as an outcome of the sporadic type of event of tenacious strain [44]. But hereditary level of concentrate more sight to discover to comprehend the physiology of diligence that could be caused by the rehashed utilization of the associated strain in sustenance-generation businesses or offices. By this method, tainted workforce, gear, or item could fill in as a vector after the rehashed presentation of a similar strain from some septic repository outside the plant [45–50]. Another examination points toward the strains separated and recognized from the outside condition inside fish slaughterhouses [46]. Different investigations have demonstrated that specialists in tainting work inside an office or with various bits of hardware may likewise be considered wellsprings of pollution [47, 49, 51] consisting of *L. monocytogenes* including locker rooms, foyers, and toilets in offices, proposing the likelihood that workforce inside the processing plant create pollution. Another report from an article proposed that the water that was used to cool the fish alongside the measuring table acted as a wellspring of defilement in enterprises [49]. It is as yet the purpose of difference in worry of the contamination, whether occasional variety has a contributing part in the detachment of *L. monocytogenes* from nourishment preparing environments [50]. Numerous examinations demonstrate no connection between regular variety and the occurrence of *L. monocytogenes* [52–54]. *L. monocytogenes* can be characterized in terms of stress tolerance. The expression "worry" in this setting is planned to mean any natural irritation that reduces the development rate (a mellow pressure) or contrarily impacts cell survival (a more serious pressure). The foundation of *L. monocytogenes* in turn around condition is the allow that makes it fit to get by finished a drawn out stretch of time in soil situations, water, mammalian and avian defecation, and in sustenance and nourishment-handling situations. It is progressively transmitted from nourishment or water assets to gastrointestinal tract of insusceptible people or creatures. With the nature of stress resilience systems, *L. monocytogenes* can hold on and act in acidic condition. It additionally reacts in conditions where there is water pressure in addition to conditions such as dry weather, low temperatures, and bile. Such a system where the pressure is getting away is controlled by a translation factor known as sigma B which is encoded by a general pressure reaction quality that connects with RNA polymerase movement which acquaint it with SigB promoter site that prompts the interpretation of sigma B protein articulation for enhancement from push [55–57]. Presently Broad Pressure Reaction (GSR) by and large very much characterized as a pressure ameliorative hereditarily for push proactive capacity.

7. Stresses encountered in food

Listeria monocytogenes can grow at temperatures as low as −0.4°C [58]. At refrigerated temperatures, they can double their number by 50 h or more [59]. When *L. monocytogenes* were treated with very-low or cold temperatures, its cell wall became more rigid, this reduced further enzymatic reaction and transportation within the cell wall [60]. It is also reported that the bacterial cell wall, due to its hardness and rigidity, the genes could express up to the mark that help to mitigate from stress conditions. Such alteration in expression of gene can involve in cell membrane function, lipid, carbohydrate, and amino acid synthesis, ribosomal structure, and biogenesis and motility [61, 62]. *L. monocytogenes* play a pivotal role in the escape from cold stress by the accumulation of low molecular weight solutes such as glycine betaine, and carnitine. Such types of biochemical solutes always present in abundance in raw or readymade food stuff which help in the survival of this pathogen at refrigerated temperature [63, 64]. The generation time of *L. monocytogenes* reduces as temperature increases by more than 20 h at 4°C in the presence of a medium of compatible solutes [59]. A new report reveals the idea of solutes that were responsible for cold stress that The BetL glycine, betaine transporter, does not seem to be involved in cryotolerance [65]. The researchers found out that exceeded expression of two genes like Gbu and OpuC were responsible for cold stress mitigation but not BetL. During that period of time, it was also observed that as the temperature increased to 8°C, the quantities of metabolite solutes like glycine, betaine, and carnitine were detected within *L. monocytogenes* as compared to 37°C [61, 66]. An account of studies examined the activity of σB with respect to cold stress escape, but the data show conflicting results. A group of researchers [61] demonstrated that some cold-induced genes were under σB control (opuCA). They manifested that these genes could be activated in σB independent manner at 4°C indicating that cold shock may be partially under σB control. They also noticed that a mutant-lacking sigB did not have reduced growth at 4°C in comparison to the wild type [61]. A new observation was made that σB gene does not participate in survival at low temperatures [67]. *L. monocytogenes* also confront the acidic environment of stomach or within the gut of the host when it comes with contaminated food sources. After entry of *L. monocytogenes* into the host through the ingestion of contaminated food, it encounters acidic conditions, first, within the stomach, and it also encounters the phagosomes after intracellular uptake. The bacterium possesses a variety of different mechanisms including the adaptive acid tolerance response (ATR), the glutamate decarboxylase (GAD) system and the arginine deaminase (ADI) system to help it overcome these acidic environments [68–70].

8. Concluding remarks

Despite the ubiquitous nature of *L. monocytogenes* in food, water, agricultural land, and cattle farms, it has harmful disadvantages. It is the cause of listeriosis in weak immune-responding person along with high risk of public health hazard. A lot of experiments and results have greatly improved the understanding of its ecology, genetics, mechanism, and physiology. By means of whole-genome sequencing method now, *L. monocytogenes* can be rapidly identified in the sources of contamination. It can greatly reduce the effort and help food producers in knowing the presence or absence of contamination in food. There is as yet noteworthy slack in our analysis and information in the worries of the exact systems that *L. monocytogenes* utilizations to detect its condition and how it couples its pressure reaction to its pathogenicity, yet the diligent work and observable research action in these fields prone to be addressed the different inquiry that still unrevealed sooner rather than later.

Acknowledgements

Authors thank the Department of Agronomy, School of Agriculture, and Lovely Professional University for offering consistent encouragement and undivided attention to the authors.

Author details

Prasann Kumar[1]* and Shweta Pathak[2]

*Address all correspondence to: prasann.21784@lpu.co.in

1 Department of Agronomy, School of Agriculture, Lovely Professional University, Jalandhar, Punjab, India

2 School of Biochemistry, Davi Ahilya University, Indore, India

References

[1] Graves LM, Helsel LO, Steigerwalt AG, Morey RE, Daneshvar MI, Roof SE, Orsi RH, Fortes ED, Milillo SR, Bakker HC, Wiedmann M, Swaminathan B, Sauders BD. *Listeria*

marthii sp. nov, isolated from the natural environment, finger lakes national forest. International Journal of Systematic and Evolutionary Microbiology. 2010;**60**:1280-1288

[2] Leclercq A, Clermont D, Bizet C, Grimont PAD, Fleche-Mateos AL, Roche SM, Buchrieser C, Cadet-Daniel V, Monnier A, Lecuit M, Allerberger F. *Listeria rocourtiae* sp. nov. International Journal of Systematic and Evolutionary Microbiology. 2010;**60**:2210-2214

[3] McLauchlin J. Human listeriosis in Britain, 1967-85, a summary of 722 cases. 2. Listeriosis in non-pregnant individuals, a changing pattern of infection and seasonal incidence. Epidemiology and Infection. 1990;**104**:191-201

[4] Schmidt-Wolf G, Seeliger HPR, Schretten-Brunner A. Mensliche listeriose. Zbl Bakt Hyg A. 1987;**265**:472-486

[5] Dee RR, Lorber B. Brain abscess due to *Listeria monocytogenes*: Case report and literature review. Review of Infectious Disease. 1986;**8**:968-977

[6] Ortel S. Listeria meningitis and septicaemia in immunocompromised patients. Acta Microbiologica et Immunologica Hungarica. 1989;**36**:153-157

[7] Halter EL, Neuhaus K, Scherer S. *Listeria weihenstephanensis* sp. nov., isolated from water plant Lemna trisulca of a German fresh water pond. International Journal of Systematic and Evolutionary Microbiology. 2012;**63**:641-647. DOI: 10.1099/ijs.0.036830-0

[8] Welshimer HJ, Donker-Voet J. *Listeria monocytogenes* in nature. Applied Microbiology. 1971;**21**:516-519

[9] Weis J, Seeliger HPR. Incidence of *Listeria monocytogenes* in nature. Applied Microbiology. 1975;**30**:29-32

[10] Dowe MJ, Jackson ED, Mori JG, Bell CR. *Listeria monocytogenes* survival in soil and incidence in agricultural soils. Journal of Food Protection. 1997;**60**:1201-1207

[11] Sauders BD, Overdevest J, Fortes E, Windham K, Schukken Y, Lembo A, et al. Diversity of Listeria species in urban and natural environments. Applied and Environmental Microbiology. 2012;**78**:4420-4433. DOI: 10.1128/AEM.00282-12

[12] Strawn LK, Fortes ED, Bihn EA, Nightingale KK, Grohn YT, Worobo RW, et al. Landscape and meteorological factors affecting prevalence of three food-borne pathogens in fruit and vegetable farms. Applied and Environmental Microbiology. 2013;**79**:588-600. DOI: 10.1128/AEM.02491-12

[13] MacGowan AP, Bowker K, McLauchlin J, Bennett PM, Reeves DS. The occurrence and seasonal changes in the isolation of Listeria spp in shop bought food stuffs, human feces, sewage and soil from urban sources. International Journal of Food Microbiology. 1994;**21**:325-334. DOI: 10.1016/0168-1605(94)90062-0

[14] Locatelli A, Depret G, Jolivet C, Henry S, Dequiedt S, Piveteau P, et al. Nation-wide study of the occurrence of *Listeria monocytogenes* in Frenchsoilsusingculture-based and molecular detection methods. Journal of Microbiological Methods. 2013;**93**:242-250. DOI: 10.1016/j.mimet.2013.03.017

[15] Wardle DA. The influence of biotic interactions on soil biodiversity. Ecology Letters. 2006;**9**:870-886. DOI: 10.1111/j.1461-0248.2006.00931

[16] Locatelli A, Spor A, Jolivet C, Piveteau P, Hartmann A. Biotic and abiotic soil properties influence survival of *Listeria monocytogenes* in soil. PloS One. 2013b;**8**:e7596. DOI: 10.1371/journal.pone.0075969

[17] Ly TMC, Muller HE. Interactions of *Listeria monocytogenes*, *Listeria seeligeri* and Listeria innocualation with protozoans. The Journal of General and Applied Microbiology. 1990;**36**:143-150. DOI: 10.2323/jgam.36.143

[18] McLaughlin HP, Casey PG, Cotter J, Gahan CGM, Hill C. Factors affecting survival of *Listeria monocytogenes* and *Listeria innocua* in soil samples. Archives of Microbiology. 2011;**193**:775-785. DOI: 10.1007/s00203-011-0716-7

[19] Watkins J, Sleath KP. Isolation and enumeration of *Listeria monocytogenes* from sewage, sewage sludge and river water. The Journal of Applied Bacteriology. 1981;**50**:1-9. DOI: 10.1111/j.1365-2672.1981.tb00865

[20] Fenlon DR, Wilson J, Donachie W. The incidence and level of *Listeria monocytogenes* contamination of food sources at primary production and initial processing. The Journal of Applied Bacteriology. 1996;**81**:641-650. DOI: 10.1111/j.13652672.1996.tb03559

[21] Nightingale KK, Schukken YH, Nightingale CR, Fortes ED, Ho AJ, Her Z, Grohn YT, McDonough PL, Wiedmann M. Ecology of transmission of *Listeria monocytogenes* infecting ruminants and in the farm environment. Applied and Environmental Microbiology. 2004;**70**:4458-4467

[22] Alghazali MR, Alazawi SK. *Listeria monocytogenes* contamination of crops grown on soil treated with seawage sludge cake. The Journal of Applied Bacteriology. 1990;**69**:642-647. DOI: 10.1111/j.1365-2672.1990.tb01557

[23] Vanrenterghem B, Huysman F, Rygole R, Verstraete W. Detection and prevalence of *Listeria monocytogenes* in the agricultural system. The Journal of Applied Bacteriology. 1991;**71**:211-217. DOI: 10.1111/j.1365-2672.1991.tb04450.x

[24] Hutchison ML, Walters LD, Moore T, Thomas DJI, Avery SM. Fate of pathogens present in livestock wastes spread onto fescue plots. Applied and Environmental Microbiology. 2005;**71**:691-696. DOI: 10.1128/AEM.71.2.691-696.2005

[25] Hutchison ML, Walters LD, Moore A, Crookes KM, Avery SM. Effect of length of time before incorporation on survival of pathogenic bacteria present in livestock wastes applied to agricultural soil. Applied and Environmental Microbiology. 2004;**70**:5111-5118. DOI: 10.1128/AEM.70.9.5111-5118.2004

[26] Jiang XP, Islam M, Morgan J, Doyle MP. Fate of *Listeria monocytogenes* in bovine manure-amended soil. Journal of Food Protection. 2004;**67**:1676-1681

[27] Steele M, Odumeru J. Irrigation water as source of foodborne pathogens on fruit and vegetables. Journal of Food Protection. 2004;**67**:2839-2849

[28] Selma MV, Allende A, Lopez-Galvez F, Elizaquivel P, Aznar R, Gil MI. Potential microbial risk factors related to soil amendments and irrigation water of potato crops. Journal of Applied Microbiology. 2007;**103**:2542-2549. DOI: 10.1111/j.1365-2672.2007.03504

[29] Ijabadeniyi OA, Debusho LK, Vanderlinde M, Buys EM. Irrigation water as a potential preharvest source of bacterial contamination of vegetables. Journal of Food Safety. 2011;**31**:452-461. DOI: 10.1111/j.1745-4565.2011.00321.x

[30] Paillard D, Dubois W, Thiebaut R, Nathier F, Hoogland E, Caumette P, et al. Occurrence of Listeria spp. in effluents of French urban wastewater treatment plants. Applied and Environmental Microbiology. 2005;**71**:7562-7566. DOI: 10.1128/AEM.71.11.7562-7566.2005

[31] Odjadjare EE, Obi LC, Okoh AI. Municipal wastewater effluents as a source of listerial pathogens in the aquatic milieu of the eastern cape province of South Africa: A concern of public health importance. International Journal of Environmental Research and Public Health. 2010;**7**:2376-2394. DOI: 10.3390/ijerph7052376

[32] Moreno Y, Ballesteros L, Garcia-Hernandez J, Santiago P, Gonzalez A, Ferrus MA. Specific detection of viable *Listeria monocytogenes* in Spanish wastewater treatment plants by fluorescent in situ hybridization and PCR. Water Research. 2011;**45**:4634-4640. DOI: 10.1016/j.watres.2011.06.015

[33] Cohan FM. Towards a conceptual and operational union of bacterial systematics, ecology, and evolution. Philosophical Transactions of the Royal Society B: Biological Sciences. 2006;**361**:1985-1996. DOI: 10.1098/rstb.2006.1918

[34] Cohan FM, Koeppel AF. The origins of ecological diversity in prokaryotes. Current Biology. 2008;**18**:R1024-R1034. DOI: 10.1016/j.cub.2008.09.014

[35] Nadon CA, Woodward DL, Young C, Rodgers FG, Wiedmann M. Correlations between molecular subtyping and serotyping of *Listeria monocytogenes*. Journal of Clinical Microbiology. 2001;**39**:2704-2707. DOI: 10.1128/JCM.39.7.27042707.2001

[36] Chenal-Francisque V, Lopez J, Cantinelli T, Caro V, Tran C, Leclercq A, et al. Worldwide distribution of major clones of *Listeria monocytogenes*. Emerging Infectious Diseases. 2011;**17**:1110-1112. DOI: 10.3201/eid/1706.101778

[37] Wiedmann M, Bruce JL, Keating C, Johnson AE, McDonough PL, Batt CA. Ribotypes and virulence gene polymorphisms suggest three distinct *Listeria monocytogenes* lineages with differences in pathogenic potential. Infection and Immunity. 1997;**65**:2707-2716

[38] Gray MJ, Zadoks RN, Fortes ED, Dogan B, Cai S, Chen Y, et al. *Listeria monocytogenes* isolates from foods and humans form distinct but overlapping populations. Applied and Environmental Microbiology. 2004;**70**:5833-5841. DOI: 10.1128/AEM.70.10.5833-5841.2004

[39] Gudmundsdottir KB, Aalbaek B, Sigurdarson S, Gunnarsson E. The diversity of *Listeria monocytogenes* strains from 10 Icelandic sheep farms. Journal of Applied Microbiology. 2004;**96**:913-921. DOI: 10.1111/j.1365-2672.2004.02183

[40] Fugett EB, Schoonmaker-Bopp D, Dumas NB, Corby J, Wiedmann M. Pulsed-field gel electrophoresis (PFGE) analysis of temporally matched *Listeria monocytogenesisolates* from human clinical cases, foods, ruminant farms, and urban and natural environments reveals source-associated as well as widely distributed PFGE types. Journal of Clinical Microbiology. 2007;**45**:865-873. DOI: 10.1128/JCM.01285-06

[41] Lunden J, Tolvanen R, Korkeala H. Acid and heat tolerance of persistent and non-persistent *Listeria monocytogenes* food plant strains. Letters in Applied Microbiology. 2008;**46**:276-280. DOI: 10.1111/j.1472-765X.2007.02305.x

[42] Ringus DL, Ivy RA, Wiedmann M, Boor KJ. Salt stress-induced transcription of σB- and CtsR-regulated genes in persistent and non-persistent *Listeria monocytogenes* strains from food processing plants. Foodborne Pathogens and Disease. 2012;**9**:198-206. DOI: 10.1089/fpd.2011.1000

[43] Magalhaes R, Ferreira V, Brandao TR, Palencia RC, Almeida G, Teixeira P. Persistent and non-persistent strains of *Listeria monocytogenes*: A focus on growth kinetics under different temperature, salt, and pH conditions and their sensitivity to sanitizers. Food Microbiology. 2016;**57**:103-108. DOI: 10.1016/j.fm.2016.02.005

[44] Ferreira V, Wiedmann M, Teixeira P, Stasiewicz MJ. *Listeria monocytogenes* persistence in food-associated environments: Epidemiology, strain characteristics, and implications for public health. Journal of Food Protection. 2014;**77**:150-170. DOI: 10.4315/0362-028X.JFP-13-150

[45] Johansson T, Rantala L, Palmu L, Honkanen-Buzalski T. Occurrence and typing of *Listeria monocytogenes* strains in retail vacuumpacked fish products and in a production plant. International Journal of Food Microbiology. 1999;**47**:111-119. DOI: 10.1016/S0168-1605(99)00019

[46] Hansen CH, Vogel BF, Gram L. Prevalence and survival of *Listeria monocytogenes* in danish aquatic and fish-processing environments. Journal of Food Protection. 2006;**69**:2113-2122

[47] Leite P, Rodrigues R, Ferreira M, Ribeiro G, Jacquet C, Martin P, et al. Comparative characterization of *Listeria monocytogenes* isolated from Portuguese farmhouse ewe's cheese and from humans. International Journal of Food Microbiology. 2006;**106**:111-121. DOI: 10.1016/j.ijfoodmicro.2005.05.017

[48] Ho AJ, Lappi VR, Wiedmann M. Longitudinal monitoring of *Listeria monocytogenes* contamination patterns in a farmstead dairy processing facility. Journal of Dairy Science. 2007;**90**:2517-2524. DOI: 10.3168/jds.2006-392

[49] Chen BY, Pyla R, Kim TJ, Silva JL, Jung YS. Incidence and persistence of *Listeria monocytogenes* in the catfish processing environment and fresh fillets. Journal of Food Protection. 2010a;**73**:1641-1650

[50] Rivoal K, Queguiner S, Boscher E, Bougeard S, Ermel G, Salvat G, et al. Detection of *Listeria monocytogenes* in raw and pasteurized liquid whole eggs and characterization by PFGE. International Journal of Food Microbiology. 2010;**138**:56-62. DOI: 10.1016/j.ijfoodmicro.2010.01.013

[51] Lomonaco S, Decastelli L, Nucera D, Gallina S, Manila Bianchi D, Civera T. *Listeria monocytogenes* in gorgonzola: Subtypes, diversity and persistence over time. International Journal of Food Microbiology. 2009;**128**:516-520. DOI: 10.1016/ j.ijfoodmicro.2008.10.009

[52] Garrec N, Picard-Bonnaud F, Pourcher AM. Occurrence of Listeria sp. and *L. monocytogenes* in sewage sludge used for land application: Effect of dewatering, liming and storage in tank on survival of Listeria species. FEMS Immunology & Medical Microbiology. 2003;**35**:275-283. DOI: 10.1016/S0928-8244(02) 00443-1

[53] Mohammed HO, Atwill E, Dunbar L, Ward T, Mcdonough P, Gonzalez R, et al. The risk of *Listeria monocytogenes* infection in beef cattle operations. Journal of Applied Microbiology. 2010;**108**:349-356. DOI: 10.1111/j.1365-2672.2009. 04446

[54] Leong D, Alvarez-Ordonez A, Jordan K. Monitoring occurrence and persistence of *Listeria monocytogenes* in foods and food processing environments in the Republic of Ireland. Frontiers in Microbiology. 2014;**5**:436. DOI: 10.3389/ fmicb.2014.00436

[55] van Schaik W, Abee T. The role of σB in the stress response of gram-positive bacteria – Targets for food preservation and safety. Current Opinion in Biotechnology. 2005;**16**:218-224. DOI: 10.1016/j.copbio.2005. 01.008

[56] Chaturongakul S, Raengpradub S, Wiedmann M, Boor KJ. Modulation of stress and virulence in *Listeria monocytogenes*. Trends in Microbiology. 2008;**16**:388-396. DOI: 10.1016/j.tim.2008.05.006

[57] O'Byrne CP, Karatzas KA. The role of sigma B (σB) in the stress adaptations of *Listeria monocytogenes*: Overlaps between stress adaptation and virulence. Advances in Applied Microbiology. 2008;**65**:115-140. DOI: 10.1016/S0065-2164(08) 00605-9

[58] Walker SJ, Archer P, Banks JG. Growth of *Listeria monocytogenes* at refrigeration temperatures. Journal of Applied Bacteriology. 1990;**68**:157-162. DOI: 10.1111/j. 1365-2672.1990.tb02561

[59] Angelidis AS, Smith GM. Role of the glycine betaine and carnitine transporters in adaptation of *Listeria monocytogenes* to chill stress in defined medium. Applied and Environmental Microbiology. 2003;**69**:7492-7498. DOI: 10.1128/AEM.69.2. 1013-1022.2003

[60] Graumann P, Maraheil MA. Some like it cold: Response of microorganisms to cold shock. Archives of Microbiology. 1996;**166**:293-300. DOI: 10.1007/ s002030050386

[61] Chan YC, Boor KJ, Wiedmann M. σB-dependent and σBindependent mechanisms contribute to transcription of *Listeria monocytogenes* cold stress genes during cold shock and cold growth. Applied and Environmental Microbiology. 2007;**73**:6019-6029. DOI: 10.1128/AEM.00714-07

[62] Cordero N, Maza F, Navea-Perez H, Aravena A, Marquez-Fontt B, Navarrete P, et al. Different transcriptional responses from slow and fast growth rate strains of *Listeria monocytogenes* adapted to low temperature. Frontiers in Microbiology. 2016;**7**:229. DOI: 10.3389/fmicb.2016.00229

[63] Zeisel SH, Mar MH, Howe JC, Holden JM. Concentrations of choline-containing compounds and betaine in common foods. The Journal of Nutrition. 2003;**133**:1302-1307

[64] Demarquoy J, Georges B, Rigault C, Royer M-C, Clairet A, Soty M, et al. Radioisotopic determination of L-carnitine content in foods commonly eaten in Western countries. Food Chemistry. 2004;**86**:137-142. DOI: 10.1016/j.foodchem. 2003.09.023

[65] Sleator RD, Gahan CGM, Hill C. A postgenomic appraisal of osmotolerance in *Listeria monocytogenes*. Applied and Environmental Microbiology. 2003;**69**:1-9. DOI: 10.1128/AEM.69.1.1-9.2003

[66] Singh AK, Ulanov AV, Li Z, Jayaswal RK, Wilkinson BJ. Metabolomes of the psychrotolerant bacterium *Listeria monocytogenes* 10403S grown at 37°C and 8°C. International Journal of Food Microbiology. 2011;**148**:107-114. DOI: 10.1016/j. ijfoodmicro.2011.05.008

[67] Utratna M, Cosgrave E, Baustian C, Ceredig RH, O'Byrne CP. Effects of growth phase and temperature on σB activity within a *Listeria monocytogenes* population: Evidence for RsbV-independent activation of σB at refrigeration temperatures. BioMed Research International. 2014;**2014**:641647. DOI: 10.1155/2014/ 641647

[68] Davis MJ, Coote PJ, O'Byrne CP. Acid tolerance in *Listeria monocytogenes*: The adaptive acid tolerance response (ATR) and growthphase-dependent acid resistance. Microbiology. 1996;**142**:2975-2982. DOI: 10.1099/ 13500872-142-10-2975

[69] Cotter PD, Gahan CG, Hill C. A glutamate decarboxylase system protects *L. monocytogenes* in gastric fluid. Molecular Microbiology. 2001;**40**:465-475. DOI: 10.1046/j.1365-2958.2001.02398

[70] Ryan S, Begley M, Gahan CG, Hill C. Molecular characterization of the arginine deiminase system in *Listeria monocytogenes*: Regulation and role in acid tolerance. Environmental Microbiology. 2009;**11**:432-445. DOI: 10.1111/j.1462-2920. 2008.01782

Section 4

Prevention and Control of Listeria monocytogenes

Contamination, Prevention and Control of *Listeria monocytogenes* in Food Processing and Food Service Environments

Frederick Tawi Tabit

Additional information is available at the end of the chapter

http://dx.doi.org/10.5772/intechopen.76132

Abstract

This chapter reviews issues related to the occurrence and growth of *Listeria monocytogenes* in food processing and food service environments. *L. monocytogenes* is a food-borne pathogen with the capacity to contaminate raw or minimally processed foods such as chilled ready-to-eat (RTE) foods. The consumption of food contaminated with *L. monocytogenes* can result in a disease known as listeriosis among vulnerable groups of people such as pregnant women and fetuses, newborns, adults between the ages of 65 and 75, and people with weakened immune systems. *L. monocytogenes* is ubiquitous and has been isolated from soil, vegetation, sewage, water, animal feed, fresh and frozen meat including poultry, slaughterhouse wastes and the feces of healthy animals and humans. The bacterium is both acid tolerant and salt tolerant. It is able to grow at refrigerator temperature, and is therefore often associated with the consumption of raw or minimally processed and often chilled RTE foods. *L. monocytogenes* is able to form biofilms on food processing and preparation surfaces, which protects it from antimicrobial action. Continuous education of vulnerable groups regarding food safety will increase their awareness of the importance of practicing safer food handling practices such as hand washing and safe storage of RTE foods as a means to prevent listeriosis.

Keywords: *Listeria monocytogenes*, ready to eat food, listeriosis, food safety

1. Introduction

Listeria monocytogenes is a bacterium which is ubiquitous in nature, and occurs frequently in food processing and handling environments [1]. The consumption of food contaminated with

L. monocytogenes can result in a disease known as listeriosis, to which pregnant women and their newborns, adults aged 65 or older, and people with weakened immune systems are particularly vulnerable [2]. In healthy adults, listeriosis is most likely to manifest as mild gastroenteritis. However, in some instances it can result in more severe symptoms, which can lead to life-threatening illnesses such as endocarditis, encephalitis or meningitis, and severe sepsis [3].

Inadequate food hygiene practices during food preparation are primarily responsible for the propagation of the bacterium and contamination of ready-to-eat (RTE) foods (**Table 1**) during processing, distribution and handling [4]. Small to medium-sized enterprises (SMEs) are more likely to experience *L. monocytogenes* outbreaks than renowned large-scale food processing enterprises owing to differences in the implementation of food safety measures [5].

RTE foods, which are often stored at low temperatures, are the type most susceptible to contamination with *L. monocytogenes* since the bacterium is psychrotrophic and possesses the ability to survive and grow in the presence of many food preservation systems, such as low pH and high salt concentrations [6]. The contamination of minimally processed fruit and vegetable products with *L. monocytogenes* is often a concern, considering that these foods, which are attractive to consumers, are often not subjected to lethal treatments during processing to inactivate potential pathogens [7]. Moreover, the manner in which RTE vegetables are sliced can affect the survival of *Listeria* and the effectiveness of decontamination procedures in the finished products. Hand tearing or manual slicing with a razor blade reduced the survival and growth of *E. coli* and *L. innocua*, probably because of less damage to the vegetable tissues and minimal leakage of nutrients from damaged plant tissues [8].

Year	Foodstuff implicated	Country of outbreak
2017	Creamy, soft, raw-milk cheeses [2]	USA
2016	Frozen vegetables [62]	USA
2016	Raw milk chocolate milk products [63]	USA
2016	Packaged salad [64]	USA
2015	Soft cheese [65]	USA
2015	Ice cream [66]	
2014	Commercially produced, pre-packaged caramel apples [67]	USA
2014	Mung bean sprouts [68]	USA
2014	Soft cheese [69]	USA
2014	Cheese products [70]	USA
2017–2017	Various food products [71]	South Africa
2017	Not determined [72]	Australia
2014	Various food products [73]	28 EU/EEA countries

CDC: Centre for Disease Control, **NICD**: National Institute for Communicable Diseases, **ECDC**: European Centre for Disease Prevention and Control, **USA**: United States of America, **EU**: European Union, **EEA**: European Economic Area.

Table 1. Some records of global ***Listeria*** outbreaks between 2014 and 2017.

L. monocytogenes has the ability to attach itself to food preparation contact surfaces and grow to form protective biofilms, which generally protect the bacterial cells from antimicrobial action during cleaning and sterilisation processes [9]. However, low concentrations (<10 μg/mL) of paenibacterin have been found to suppress the growth of *L. monocytogenes* within the biofilm matrix as well as to down-regulate the genes involved in biofilm formation [10]. Considering that *L. monocytogenes* is a food-borne pathogen of public interest [11], the objective of this paper is to review issues related to the occurrence and growth of *L. monocytogenes* in food processing and food service environments.

2. Health and economic impacts of listeriosis

Globally, billions of people are at risk every year and thousands die as a result of consuming unsafe food [12]. In the United States of America (USA), listeriosis has been identified as the third leading cause of death from food-borne illness, after non-typhoidal *Salmonella* and *Toxoplasma gondii*, despite its rarity [13]. In Africa, food-borne illness continues to be a major health threat, especially for vulnerable groups such as infants, pregnant women and their newborns as well as immune-compromised individuals such as elderly people and those with HIV/AIDS [14].

In humans, invasive listeriosis is characterized by septicemia, meningitis, and abortion in pregnant women [15]. Listeriosis in pregnant women can result in premature labor, stillbirth, abortion, and neonatal infection, with high neonatal mortality [16]. It should be noted that *L. monocytogenes* infection in healthy individuals does not necessarily result in invasive disease. The incubation period of listeria-related gastroenteritis can range from 1 to 24 days, but the average incubation period has been found to be less than 24 hours. After the incubation period, prominent symptoms will include fever, then diarrhea, arthralgia, myalgia, and headache. Other common symptoms are nausea, vomiting, abdominal pain and watery diarrhea. In healthy individuals, the illness tends to last between 1 and 3 days, with a very low rate of hospitalization [17].

Listeriosis may have an economic impact in the form of costs incurred by the government in funding health institutions to deal with the problem [18]. Other costs can take the form of legal costs emanating from lawsuits imposed on food production companies arising from illness and death due listeriosis [19].

3. Ecology and growth conditions of *Listeria monocytogenes* in the food chain

L. monocytogenes are ubiquitous bacteria that can be found in different environments such as soil and water, and especially in food-manufacturing environments [20]. Many *Listeria* species have been isolated from soil, vegetation, sewage, water, animal feed, fresh and frozen meat including poultry, slaughterhouse wastes and the feces of healthy animals, including humans [21]. Animals have been found to be carriers of *L. monocytogenes*, hence the contamination of foods of animal origin, such as meats and dairy products [22].

L. monocytogenes can survive a low pH of 5.5 through a phenomenon known as the acid tolerance response (ATR), which causes cells to be more resistant in adverse acidic conditions [23]. The bacterium, which is notable for its persistence in food-manufacturing environments, is relatively salt-tolerant and is able to grow at refrigerator temperature, and is therefore often associated with the consumption of raw or minimally processed and often chilled RTE foods (e.g., soft and semi-soft cheese and smoked fish products), which are consumed without further processing [24, 25].

4. The occurrence of *L. monocytogenes* in the food processing environment

L. monocytogenes is able to attach to food processing surfaces and multiply to form biofilms in inaccessible locations in processing facilities [9]. Biofilms protect the bacterium against antimicrobial action, enabling it to colonize food processing equipment, conveyor belts, pipes, floors and drainage systems and to persist for months or even years, cross-contaminating different surfaces in food processing plants [26]. The formation of biofilms on various food contact surfaces by *L. monocytogenes* makes it extremely difficult to control this pathogen effectively, especially in processing plants where inadequate cleaning has been carried out [27].

5. The occurrence of *L. monocytogenes* in RTE foods and food contact surfaces in food service facilities

RTE foods have gained considerable popularity in many developing and developed countries because of their perceived better flavor, affordability and accessibility [28]. However, numerous *L. monocytogenes* outbreaks have been associated with RTE foods [29]. The prevalence of *L. monocytogenes* in RTE food is a major concern relating to food safety because RTE foods are consumed without further processing (cooking) or washing at home (**Table 2**). It is for this reason that stringent microbiological guidelines need to be formulated and followed to ensure that processors produce RTE food that is safe [30]. Implicated RTE foods include RTE deli meats, raw milk and other raw milk dairy products (soft cheese) (**Table 1**). Between 1999 and 2011, 73% of all food-borne outbreaks of listeriosis that occurred in the United Kingdom (UK) were attributed to the consumption of sandwiches [31].

The presence of *L. monocytogenes* in RTE food is attributed to contamination during production, distribution or storage [32]. *L. monocytogenes* contamination in various food factory environments has been reported at nearly all stages of processing ([5] and Rodrigues et al. [33]). When compared with other food-borne pathogens such as *Staphylococcus aureus*, *E. coli* 015:H7, and *Salmonella* and *Shigella* species, *L. monocytogenes* has been found to be most prevalent on food contact surfaces in food service establishments [34]. Owing to its ability to grow in contaminated food during storage at refrigeration temperature, *L. monocytogenes* has

Food groups	Susceptible food products
Meat	Processed meat products such as ground beef, sausages, deli ham, beef hot dogs and meat-related sandwich products (e.g., pork, beef)
Poultry	Processed chicken such as deli chicken, deli turkey, eggs, and related sandwich products
Fish	Cooked shrimps, sushi, smoked salmon, seafoods, and related sandwich and salad dishes
Dairy	Cheese, yogurt
Fruit and vegetables	Cabbage, lettuce, cucumber, frozen green beans, peanut butter, vegetable salads, raw sprouts, cantaloupe melon and related salad dishes
Cereal and baked products	Pasta, cakes, pies, sausage rolls

Table 2. Listeria in food: foods that are susceptible to contamination by ***Listeria monocytogenes*** [57, 74].

been found in raw and processed RTE foods that required low temperature storage [35]. The high volume of food products such as meat, vegetables, dairy products and fruits that pass through the cold chain in food service establishments could contribute to the high incidence of *L. monocytogenes* in RTE food and on food contact surfaces [34, 36].

Inadequate cleaning procedures and hygiene practices can promote the formation of biofilms on food contact surfaces in food service establishments, thereby increasing the chances of *L. monocytogenes* cross-contamination within food service facilities [37]. Because biofilms are able to resist most sanitisers and disinfectants used, cross-contamination by *L. monocytogenes* poses a serious food safety risk in food service establishments, including domestic kitchens [38]. The ease with which *L. monocytogenes* is able to adhere to food contact surfaces and form biofilms increases the likelihood of its persisting on food contact surfaces, and hence cross-contaminating the final food products [39, 40]. The presence of food debris on food contact surfaces encourages the formation of *L. monocytogenes* biofilms [41].

6. The occurrence of *L. monocytogenes* on food contact surfaces in domestic kitchens

Inadequate hygiene practices in domestic kitchens may contribute to the persistence of food-borne pathogens, thereby compromising the safety of foods produced there [42]. Home kitchens have been found to be a significant location where food-borne illnesses are acquired. A survey conducted in the domestic kitchens of consumers aged 60 and above in the UK indicated that a large number of foods in home refrigerators were beyond the use-by date and up to 66% of opened RTE foods had been stored beyond the recommended 2 days after opening [43]. A study of the occurrence of *Listeria* spp. on food contact surfaces in domestic kitchens in the Netherlands found high levels of *L. monocytogenes* on dish-cloths and in bathrooms, but low levels on kitchen sinks, washing-up brushes and refrigerators [44]. Many researchers have found high levels of *L. monocytogenes* on refrigerator surfaces in domestic kitchens [45].

7. Legislation relating to the occurrence of *Listeria monocytogenes* in foods

Most food legislation stipulates the microbial criteria for food-borne bacteria such as *L. monocytogenes* or their toxins and metabolites in specific foods. These criteria often prescribe the acceptable levels of these bacteria or their toxins in food products available on the market [46]. Most foods that support the growth of *L. monocytogenes* should be the focus of risk management efforts. Countries such as Germany, the Netherlands and France have set a tolerance level of 100 colony forming units (cfu) of *L. monocytogenes* per gram of food at the time of consumption while others, such as the USA and Italy, require a total absence of *L. monocytogenes* in 25 g of food [47]. The new criteria for *L. monocytogenes* in RTE food gazetted by Food Standards Australia-New Zealand on 31 July 2014 prescribe two sets of criteria for *L. monocytogenes* for application based on whether the growth of the bacterium does or does not occur inherently in a particular RTE food. These criteria include fewer than 100 cfu of *L. monocytogenes* per gram of food in which the growth of *L. monocytogenes* is not likely to occur, and that *L. monocytogenes* should not be detected in 25 g of food in which the growth of *L. monocytogenes* is likely to occur [48].

The Food Safety Standard of Ireland has prescribed the following in relation to *L. monocytogenes*: *L. monocytogenes* should be absent in 25 g of RTE food destined for infant consumption or for serving as a special food for medical purposes in up to 10 collected food samples. Similarly, in the case of RTE foods that are able to support the growth of *L. monocytogenes*: *L. monocytogenes* should be absent in 25 g of RTE food following production or should not exceed 100 cfu per gram of food placed on the market during its shelf life, in up to 5 collected food samples. Lastly, in the case of RTE foods that are not able to support the growth of *L. monocytogenes*: *L. monocytogenes* should not exceed 100 cfu per gram of food placed on the market during its shelf life, in up to 5 collected food samples [49].

8. Methods commonly used for the identification of *L. monocytogenes* in foods

8.1. Culture methods

L. monocytogenes can be isolated from contaminated samples by subjecting them to pre-enrichment. This entails mixing samples with enrichment media such as *Listeria* Enrichment Broth (Sigma), after which the enrichment samples can be cultured on *L. monocytogenes*-specific agar plates such as Listeria Mono Differential Agar (Sigma). Isolation can be performed using various other media and procedures [50]. Thereafter, pure cultures of *L. monocytogenes* to be used for downstream identification and characterization analysis can be prepared by isolating individual colonies from agar plates [51]. The culture-based methods are often used in combination with immunoassay- or molecular PCR-based methods for accurate detection of *L. monocytogenes* in food samples [52].

8.2. Immunoassay

During immunoassay, monoclonal antibodies specific to *L. monocytogenes* can be incorporated into various techniques for identification. Immunoassay tests usually have high specificity and are fast and easy to use, but do not permit identification to species level. Another disadvantage of this method is that the presence of a low number of listeria cells in a sample can give rise to a false positive [53]. Various variants of immunoassays are available, including sandwich-type enzyme-linked immunosorbent assay (S-ELISA) [54], nanoparticle immunoassay [55], and enzyme-linked fluorescent assay (ELFA) [56].

8.3. PCR-related methods

PCR-based techniques involve the amplification of a specific gene segment of *L. monocytogenes* such as HlyA-, Iap-, PrfA and SsrA using specific primers followed by monitoring of the amplified segment using agarose gel electrophoresis or other detection techniques such as SYBR Green [57]. Similarly, the 16S rRNA genes of *L. monocytogenes* can be amplified, sequenced, and searched against existing databases for identification [52]. The disadvantage of PCR-based techniques is related to the costs associated with the purchase of the instrument and reagent, as well as the expertise required to conduct the experiments [58].

9. Prevention and control of *Listeria monocytogenes* in food systems

The prevention and control of *L. monocytogenes* in RTE foods is paramount in protecting consumers against listeriosis. In a document entitled "Guidelines on the application of general principles of food hygiene to the control of *L. monocytogenes* in foods" the World Health Organization has provided guidelines that can be followed to minimize the likelihood of the occurrence of *L. monocytogenes* in RTE foods. According to the [59], food safety measures need to be carried out at different levels of a food production environment, and must include:

1. Establishing the design and adequacy of a production facility: proper location and layout, and adequate equipment and facilities such as water supply, drainage, toilets, temperature control, storage, and hand washing basins.
2. Control of food safety hazards and implementation of hygiene practices throughout the food production line. Accredited HACCP implementation programme.
3. Establishment of adequate sanitary conditions and maintenance of the production facilities; effective cleaning programmes; pest control and proper waste management; and effective monitoring of cleaning programmes.
4. Ensuring adequate implementation of personal hygiene, health status, personal cleanliness and personal behavior of staff.
5. Ensuring adequate and properly functioning transport facilities; these should be well maintained and fit for purpose.
6. Continuous training of staff working in the food production environment, including refresher training.

While the food industry is taking numerous measures to protect foods from *Listeria*, consumers of RTE food, especially those belonging to the vulnerable groups, must take suitable

Vulnerable consumer group	Reason for vulnerability	Recommended preventive food hygiene measures
Pregnant women	Weak immune system due to hormonal changes	• Wash and dry your hands before and after touching and preparing ready-to-eat food.
Unborn fetuses and newborn babies	Undeveloped immune system	• Refrigerator food contact surfaces should be clean and sanitized regularly, and operate below 5°C.
People over the age of 65	Weak immune system due to ageing	• Kitchen utensils such as knives, cutting boards and graters must be washed before and after being used in preparing ready-to-eat foods.
People with diseases such as cancer, leukemia, AIDS, diabetes, or liver or kidney disease	Weak immune system due to disease	• Minimally processed fruits and vegetables must be washed thoroughly in flowing water prior to consumption.
People on drugs that can suppress the immune system such prednisone or cortisone	Suppressed immune system due to drugs	• Store raw meat separately from and below cooked and ready-to-eat food in the refrigerator. • Protein-rich foods containing meat, fish, chicken, egg, sprouts and dairy foods that have cooled to room temperature must be discarded.
People undergoing organ transplant	Suppressed immune system due to drug administration	• Protein-rich foods must be kept either hot (60°C or hotter) or cold (5°C or colder).

Table 3. Listeria in food: Advice to people vulnerable to listeriosis [75].

precautions during the handling of food in their households to prevent the growth and contamination of food by *L. monocytogenes* (**Table 3**).

10. Consumer awareness of listeriosis

Continuous provision of food safety education to consumers through various channels such as social media increases consumer awareness of the need for safer food handling practices such as hand washing and safe storage of RTE food [60]. The food standard agency of the UK has identified and targeted consumers who are at risk of contracting listeriosis. Vulnerable people, many of whom live and obtain their food independently include those with various forms of cancer, diabetes, alcoholism and diseases of the kidneys, liver, cardiovascular system (e.g., heart disease), digestive system (e.g., Crohn's disease) and musculoskeletal/connective tissue system (e.g., lupus) [61]. Even though most consumers of food sold by street vendors may not have confidence in the safety of RTE foods sold on the street, this often does not affect their preference for such foods because of their affordability, availability and convenience [28].

11. Conclusions

RTE foods have gained considerable popularity in many developing and developed countries because of their perceived better flavor, affordability and convenience. Consumers will continue to consume RTE foods despite their association with *L. monocytogenes* outbreaks. While

http://dx.doi.org/10.5772/intechopen.76132

most food processing industries are taking measures to protect foods from listeria, consumers of RTE food, especially those belonging to vulnerable groups, must take suitable precautions during the handling of food to prevent the growth of *L. monocytogenes* and contamination of food by this organism. Continuous identification of those groups of consumers vulnerable to listeriosis and food safety education directed at them specifically will increase their awareness of the need for safer food handling practices such as hand washing and safe storage of RTE food in an effort to prevent listeriosis.

Acknowledgements

I would like to acknowledge my wife, Wendy Tabit, for reading this manuscript and making suggestions.

Conflict of interest

I declare that I have no conflict of interest regarding the publication of this research chapter.

Author details

Frederick Tawi Tabit

Address all correspondence to: tabitft@uinisa.ac.za

University of South Africa, Johannesburg, South Africa

References

[1] Madden RH, Hutchison M, Jordan k, Pennone V, Gundogdu O, Corcionivoschi N. Prevalence and persistence of *Listeria monocytogenes* in premises and products of small food business operators in Northern Ireland. Food Control. 2018;**87**:70-78

[2] Centre for Disease Control and Prevention (CDC). *Listeria* (listeriosis): Multistate Outbreak of Listeriosis Linked to Soft Raw Milk Cheese Made by Vulto Creamery (Final Update) [Internet]. 2017. Available from: https://www.cdc.gov/listeria/outbreaks/soft-cheese-03-17/index.html [Accessed: February 12, 2018]

[3] Roberts AJ, Wiedmann M. Pathogen, host and environmental factors contributing to the pathogenesis of listeriosis. Cellular and Molecular Life Sciences. 2003;**60**(5):904-918

[4] Beno SM, Stasiewicz MJ, Andrus AD, Ralyea RD, Kent DJ, Martin NH, Wiedmann M, Boor KJ. Development and validation of pathogen environmental monitoring programs for small cheese processing facilities. Journal of Food Protection. 2016;**79**(12):2095-2106

[5] Leong D, NicAogáin K, Luque-Sastre L, McManamon O, Hunt K, Alvarez-Ordóñez A, Scollard O, Schmalenberger A, Fanning S, O'Byrne C, Jordan K. A 3-year multi-food study of the presence and persistence of *Listeria monocytogenes* in 54 small food businesses in Ireland. International Journal of Food Microbiology. 2017;**249**:18-26

[6] Ryan S, Hill C, Gahan CGM. Acid stress responses in *Listeria monocytogenes*. Advances in Applied Microbiology. 2008;**65**:67-91

[7] Oliveira MAD, Maciel de Souza V, Morato Bergamini AM, De Martinis ECP. Microbiological quality of ready-to-eat minimally processed vegetables consumed in Brazil. Food Control. 2011;**22**(8):1400-1403

[8] Gleeson E, O'Beirne D. Effects of process severity on survival and growth of *Escherichia coli* and *Listeria innocua* on minimally processed vegetables. Food Control. 2005;**16**(8): 677-685

[9] Kocot AM, Olszewska MA. Biofilm formation and microscopic analysis of biofilms formed by *Listeria monocytogenes* in a food-processing context. LWT - Food Science and Technology. 2017;**84**:47-57

[10] Li R, Du W, Yang J, Liu Z, Yousef AE. Control of *Listeria monocytogenes* biofilm by paenibacterin, a natural antimicrobial lipopeptide. Food Control. 2018;**84**:529-535

[11] Drevets DA, Jelinek TA, Freitag NE. *Listeria monocytogenes*-infected phagocytes can initiate central nervous system infection in mice. Applied and Environmental Microbiology. 2001;**69**(3):1344-1350

[12] Quinlan JJ. Foodborne illness incidence rates and food safety risks for populations of low socioeconomic status and minority race/ethnicity: A review of the literature. International Journal of Environmental Research Public Health. 2013;**10**:3634-3652

[13] Scallan E, Hoekstra RM, Angulo FJ, Tauxe RV, Widdowson M-A, Roy SL, Jones JL, Griffin PM. Foodborne illness acquired in the United States – Major pathogens. Emerging Infectious Diseases. 2011;**17**(1):7-15

[14] Mensah P, Mwamakamba L, Mohammed C, Nsuemilang D. Public health and food safety in the WHO African region. African Journal of Food, Agriculture, Nutrition, and Development. 2012;**12**(4):6317-6335

[15] Norton DM, Braden CR. Foodborne listeriosis. In: Ryser EH, Marth EH, editors. *Listeria*, Listeriosis and Food Safety. 3rd ed. Boca Raton: CRC Press Taylor & Francis Group; 2007. pp. 305-356

[16] Mylonakis E, Paliou M, Hohmann EL, Calderwood SB, Wing EJ. Listeriosis during pregnancy: A case series and review of 222 cases. Medicine (Baltimore). 2002;**81**:260-269

[17] McNeill C, Sisson W, Jarrett A. Listeriosis: A resurfacing menace. The Journal for Nurse Practitioners. 2017;**13**(10):647-654

[18] Scharff LR. Economic burden from health losses due to foodborne illnesses in the United States. Journal of Food Protection. 2012;**75**(1):123-131

[19] Thomas MK, Vriezen R, Farber JM, Currie A, Schlech W, Fazil A. Economic cost of a Listeria monocytogenes outbreak in Canada, 2008. Foodborne Pathogens and Disease. 2015;**12**(12):966-971

[20] Hamon M, Bierne H, Cossart P. *Listeria monocytogenes*: A multifaceted model. Nature Reviews Microbiology. 2006;**4**(6):423-434

[21] Gebretsadik S, Kassa T, Alemayehu H, Huruy K, Kebede N. Isolation and characterization of *Listeria monocytogenes* and other *Listeria* species in foods of animal origin in Addis Ababa, Ethiopia. Journal of Infection and Public Health. 2011;**4**(1):22-29

[22] Dowe MJ, Jackson ED, Mori JG, Bell CR. *Listeria monocytogenes* survival in soil and incidence in agricultural soils. Journal of Food Protection. 1997;**60**:1201-1207

[23] Gandhi M, Chikindas M. Review: Listeria: A foodborne pathogen that knows how to survive. International Journal of Food Microbiology. 2007;**113**:1-15

[24] Borucki MK, Peppin JD, White D, Loge F, Call DR. Variation in biofilm formation among strains of *Listeria monocytogenes*. Applied and Environmental Microbiology. 2003;**69**: 7336-7342

[25] Misiou O, van Nassau TJ, Lenz CA, Vogel RF. The preservation of *Listeria*-critical foods by a combination of endolysin and high hydrostatic pressure. International Journal of Food Microbiology. 2018;**266**(2):355-362

[26] Berrang ME, Frank JF, Meinersmann RJ. *Listeria monocytogenes* biofilm formation on silver ion impregnated cutting boards. Food Protection Trends. 2010;**30**:168-171

[27] Belessi CEA, Gounadaki AS, Psomas AN, Skandamis PN. Efficiency of different sanitation methods on *Listeria monocytogenes* biofilms formed under various environmental conditions. International Journal of Food Microbiology. 2011;**145**:S46-S52

[28] Asiegbu CV, Lebelo SL, Tabit FT. The food safety knowledge and microbial hazards awareness of consumers of ready-to-eat street-vended food. Food Control. 2016;**60**:422-429

[29] Lokerese RFA, Maslowska-Corker KA, van de Wardt LC, Wijtzes T. Growth capacity of *Listeria monocytogenes* in ingredients of ready-to-eat salads. Food Control. 2016;**60**:338-345

[30] Mika-Krajnik M, Yuk H, Kumar A, Yang Y, Zheng Q, Kim M, Ghate V, Yuan W, Pang X. Ensuring food security through enhancing microbiological food safety. Cosmos. 2015;(01):69-87

[31] Little CL, Amar CFL, Awofisayo A, Grant KA. Hospital-acquired listeriosis associated with sandwiches in the UK: A cause for concern. Journal of Hospital Infection. 2012;**82**(1): 13-18

[32] Lambertz ST, Nilsson C, Brådenmark A, Sylvén S, Johansson A, Jansson LM, Lindblad M. Prevalence and level of *Listeria monocytogenes* in ready-to-eat foods in Sweden 2010. International Journal of Food Microbiology. 2012;**160**(1):24-31

[33] Rodrigues C, de Sá C, de Melo C. An overview of *Listeria monocytogenes* contamination in ready to eat meat, dairy and fishery foods. Ciência Rural. 2017;**47**(2):1-8

[34] Sibanyoni JJ. Food safety and quality assurance measures of the national school nutrition programme in Mpumalanga Province, South Africa. [Phd Dissertation]: University of South Africa; 2017

[35] Du X-J, Zhang X, Wang X-Y, Su Y-L, Li P, Wang S. Isolation and characterization of *Listeria monocytogenes* in Chinese food obtained from the central area of China. Food Control. 2017;**74**:9-16

[36] Stepanović S, Dakić I, Martel A, Vaneechoutte M, Morrison D, Shittu A, Ježek P, Decostere A, Devriese LA, Haesebrouck F. A comparative evaluation of phenotypic and molecular methods in the identification of members of the *Staphylococcus sciuri* group. Systematic and Applied Microbiology. 2005;**28**(4):353-357

[37] Carpentier B, Cerf O. Review-persistence of *Listeria monocytogenes* in food industry equipment and premises. International Journal of Food Microbiology. 2011;**145**:1-8

[38] Aureli P, Fiorucci GC, Caroli D, Marchiaro G, Novara O, Leone L, Salmaso S. An outbreak of febrile gastroenteritis associated with corn contaminated by *Listeria monocytogenes*. New England Journal of Medicine. 2000;**342**(17):1236-1241

[39] Di Bonaventura G, Piccolomini R, Paludi D, D'Orio V, Vergara A, Conter M, Ianieri A. Influence of temperature on biofilm formation by *Listeria monocytogenes* on various food-contact surfaces: Relationship with motility and cell surface hydrophobicity. Journal of Applied Microbiology. 2008;**104**(6):1552-1561

[40] Lourenço A, Rego F, Brito L, Frank JF. Evaluation of methods to assess the biofilm-forming ability of *Listeria monocytogenes*. Journal of Food Protection. 2012;**75**(80):1411-1417

[41] Blackman IC, Frank JF. Growth of *Listeria monocytogenes* as a biofilm on various food-processing surfaces. Journal of Food Protection. 1996;**59**(8):827-831

[42] Catellani P, Scapin RM, Alberghini L, Radu IL, Giaccone V. Levels of microbial contamination of domestic refrigerators in Italy. Food Control. 2014;**42**:257-262

[43] Evans EW, Redmond EC. Analysis of older adults' domestic kitchen storage practices in the United Kingdom: Identification of risk factors associated with Listeriosis. Journal of Food Protection. 2015;**78**(4):738-745

[44] Beumer RR, Te Giffel MC, Spoorenberg E, Rombouts FM. *Listeria* species in domestic environments. Epidemiology and Infection. 1996;**117**(3):437-442

[45] Azevedo I, Regalo M, Mena C, Cameiro L, Teixera P, Hogg T, Gibbs PA. Incidence of *Listeria* spp. in domestic refrigerators in Portugal. Food Control. 2005;**16**(2):121-124

[46] Luber P. The codex Alimentarius guidelines on the application of general principles of food hygiene to the control of *Listeria monocytogenes* in ready-to-eat foods. Food Control. 2011;**22**(9):1482-1483

[47] Nørrung B. Microbiological criteria for *Listeria monocytogenes* in foods under special consideration of risk assessment approaches. International Journal of Food Microbiology. 2000;**62**(3):217-221

[48] Food Standards Australia–New Zealand (FSANZ). Criteria for *Listeria monocytogenes* – Microbiological Limits for Foods [Internet]. 2014. Available from: http://www.foodstandards.gov.au/code/proposals/Documents/P1017-MicroAppR-SD2.pdf [Accessed: February 12, 2018]

[49] Food Safety Authority of Ireland: *Listeria monocytogenes*. Microbial factsheet series Issue No 1 September 2011 [Internet]. 2016. Available from: https://www.fsai.ie/listeriamonocytogenes.html [Accessed: February 13, 2018]

[50] Rosimin AA, Kim M-J, Joo I-S, Suh S-H, Kim KS. Simultaneous detection of pathogenic *Listeria* including atypical *Listeria innocua* in vegetables by a quadruplex PCR method. LWT - Food Science and Technology. 2016;**69**:601-607

[51] Dwivedi HP, Jaykus L. Detection of pathogens in foods: The current state-of-the-art and future directions. Critical Reviews in Microbiology. 2011;**37**(1):40-63

[52] Liu H, Lu L, Pan Y, Sun X, Hwang C-A, Zhao Y, Wu VCH. Rapid detection and differentiation of *Listeria monocytogenes* and *Listeria* species in deli meats by a new multiplex PCR method. Food Control. 2015;**52**:78-84

[53] Capita R, Alonso-Calleja C, Moreno B. García-Fernández MC: Occurrence of *Listeria* species in retail poultry meat and comparison of a cultural/immunoassay for their detection. International Journal of Food Microbiology. 2001;**65**(1-2):75-82

[54] Liu A, Xiong Q, Shen L, Li W, Zeng Z, Li C, Liu S, Liu Y, Han G. A sandwich-type ELISA for the detection of *Listeria monocytogenes* using the well-oriented single chain Fv antibody fragment. Food Control. 2017;**79**:156-161

[55] Jaakohuhta S, Härmä H, Tuomola M, Lövgren T. Sensitive *Listeria* spp. immunoassay based on europium(III) nanoparticulate labels using time-resolved fluorescence. International Journal of Food Microbiology. 2007;**114**(3):288-294

[56] Sewell AM, Warburton DW, Boville A, Daley EF, Mullen K. The development of an efficient and rapid enzyme linked fluorescent assay method for the detection of *Listeria* spp. from foods. International Journal of Food Microbiology. 2003;**81**(2):123-129

[57] Cheng J-Q, Healey S, Regan P, Laksanalamai P, Hu Z. PCR-based methodologies for detection and characterization of *Listeria monocytogenes* and *Listeria ivanovii* in foods and environmental sources. Food Science and Human Wellness. 2017;**6**(2):39-59

[58] Tabit FT. Advantages and limitations of potential methods for the analysis of bacteria in milk: A review. Journal of Food Science and Technology. 2016;**53**(1):42-49

[59] Codex Alimentarius: Guidelines on the application of general principles of food hygiene to the control of *Listeria monocytogenes* in foods, CAC/GL 61-2007 [Internet]. 2007. Available from: http://www.fao.org/input/download/standards/10740/CXG_061e.pdf [Accessed: February 12, 2018]

[60] Lin C-TJ, Jensen KL, Yen ST. Awareness of foodborne pathogens among US consumers. Food Quality and Preference. 2005;**16**(5):401-412

[61] Food Standard Agency (FSA): *Listeria* guidance for healthcare and social care organisations [Internet]. 2016. Available from: https://www.food.gov.uk/sites/default/files/listeria-guidance-june2016-rev.pdf [Accessed: February 14, 2018]

[62] CDCa. *Listeria* (listeriosis): Multistate Outbreak of Listeriosis Linked to Frozen Vegetables (Final Update) [Internet]. 2016. Available from: https://www.cdc.gov/listeria/outbreaks/frozen-vegetables-05-16/index.html [Accessed: February 12, 2018]

[63] CDCb. *Listeria* (listeriosis): Multistate Outbreak of Listeriosis Linked to Raw Milk Produced by Miller's Organic Farm in Pennsylvania (Final Update) [Internet]. 2016. https://www.cdc.gov/listeria/outbreaks/raw-milk-03-16/index.html [Accessed: February 12, 2018]

[64] CDCc. *Listeria* (listeriosis): Multistate Outbreak of Listeriosis Linked to Packaged Salads Produced at Springfield, Ohio Dole Processing Facility (Final Update) [Internet]. 2016. Available from: https://www.cdc.gov/listeria/outbreaks/bagged-salads-01-16/index.html [Accessed: February 12, 2018]

[65] CDCa. *Listeria* (listeriosis): Multistate Outbreak of Listeriosis Linked to Soft Cheeses Distributed by Karoun Dairies, Inc. (Final Update) [Internet]. 2015. Availlable from: https://www.cdc.gov/listeria/outbreaks/soft-cheeses-09-15/index.html [Accessed: February 12, 2018]

[66] CDCb. *Listeria* (listeriosis): Multistate Outbreak of Listeriosis Linked to Blue Bell Creameries Products (Final Update) [Internet]. 2015. Available from: https://www.cdc.gov/listeria/outbreaks/ice-cream-03-15/index.html [Accessed: February 12, 2018]

[67] CDCa. *Listeria* (listeriosis): Multistate Outbreak of Listeriosis Linked to Commercially Produced, Prepackaged Caramel Apples Made from Bidart Bros. Apples (Final Update) [Internet]. 2014. Available from: https://www.cdc.gov/listeria/outbreaks/caramel-apples-12-14/index.html [Accessed: February 12, 2018]

[68] CDCb. *Listeria* (listeriosis): Wholesome Soy Products, Inc. Sprouts and Investigation of Human Listeriosis Cases (Final Update) [Internet]. 2014. Available from: https://www.cdc.gov/listeria/outbreaks/bean-sprouts-11-14/index.html [Accessed: February 12, 2018]

[69] CDCc. *Listeria* (listeriosis): Oasis Brands, Inc. Cheese Recalls and Investigation of Human Listeriosis Cases (Final Update) [Internet]. 2014. Available from: https://www.cdc.gov/listeria/outbreaks/cheese-10-14/index.html [Accessed: February 12, 2018]

[70] CDCd. *Listeria* (listeriosis): Multistate Outbreak of Listeriosis Linked to Roos Foods Dairy Products (Final Update) [Internet] 2014. Available from: https://www.cdc.gov/listeria/outbreaks/cheese-02-14/index.html [Accessed: February 12, 2018]

[71] National Institute of Communicable Diseases, (NICD): Situation report on listeriosis outbreak, South Africa, 2017 [Internet]. 2017. Available from: http://www.nicd.ac.za/wp-content/uploads/2017/12/NICD_Situation_report_on_listeriosis_outbreak_South_Africa_04_December_2017.pdf [Accessed: February 16, 2018]

[72] Herald Sun: One dead in Victorian listeria food poisoning surge [Internet]. 2017. Available from: http://www.heraldsun.com.au/news/victoria/one-victorian-dead-in-listeria-food-poisoning-surge/news-story/16ca34ccf916371d3b05695df8b296ec [Accessed: February 17, 2018]

[73] European Centre for Disease Prevention and Control (ECDC): Annual Epidemiological Report 2016 – Listeriosis. Stockholm: ECDC; 2016. [Internet]. 2016. Available from: http://ecdc.europa.eu/sites/portal/files/documents/Listeriosis%20-%20Annual%20epidemiological%20report_0.pdf [Accessed: February 15, 2018]

[74] Hamidiyan N, Salehi-Abargouei A, Rezaei Z, Tafti RD, Akrami-Mohajeria F. The prevalence of *Listeria* spp. food contamination in Iran: A systematic review and meta-analysis. Food Research International. DOI: 10.1016/j.foodres.2018.02.038

[75] Food Standards Australia–New Zealand (FSANZ): Listeria and food – advice for people at risk [Online]. 2018. Available from: http://www.foodstandards.gov.au/consumer/safety/listeria/documents/listeria-1.pdf [Accessed: February 15, 2018]

Section 5

The Impact of Environmental Stresses and the Virulence of Listeria monocytogenes

Chapter 7

The Impact of Environmental Stresses in the Virulence Traits of *Listeria monocytogenes* Relevant to Food Safety

Sofia Araújo Pereira, Ângela Alves,
Vânia Ferreira and Paula Cristina Maia Teixeira

Additional information is available at the end of the chapter

http://dx.doi.org/10.5772/intechopen.76287

Abstract

Listeria monocytogenes is a foodborne pathogen, which causes listeriosis disease among humans and other animal species. Infections in humans mainly occur in immunocompromised individuals and are caused by the consumption of ready-to-eat and raw food products contaminated with the pathogen. To ensure survival in nature, *L. monocytogenes* easily adapts to different environmental conditions, and that justifies the hurdles to prevent bacterial growth inside the food chain. Exposure to a single or multiple sublethal stresses, as those impaired by food processing, food matrices, and the gastrointestinal tract, can enhance tolerance of *L. monocytogenes* to stresses and increase its survival and pathogenesis. This chapter summarizes the current information on the adaptive response of *L. monocytogenes* to different stresses, namely (1) cold stress, (2) acid stress, (3) osmotic stress, (4) desiccation stress, and (5) high hydrostatic pressure, and the impact of these stresses on *L. monocytogenes* virulence. The objective is to provide the background information that is necessary for the development of scientifically sound control strategies to improve food safety and to reduce the uncertainty of microbial risk assessments, associated to limited knowledge on the behavior of cells capable to adapt and survive stresses.

Keywords: *Listeria monocytogenes*, stress response, virulence

1. Introduction

Listeria monocytogenes is a pathogenic bacterium capable of causing listeriosis disease in humans and other animals. *L. monocytogenes* has a ubiquitous distribution in the environment [1].

Human listeriosis is on the top five most commonly reported zoonosis under the surveillance of the European Union (EU) and presents the highest case fatality rate, that is, 16.2% [2]. The incidence of invasive forms of the disease is higher in risk groups, such as the elderly, immunocompromised individuals, pregnant women, and newborns. In countries with established surveillance programs, the incidence of listeriosis is reported to be increasing, and the distribution of cases is shifting, primarily affecting elderly persons. In 2016, most cases of listeriosis were reported in individuals over 64 years of age [2]. This is worrisome, as advances in the field of medicine are leading to growing life expectancies; therefore, an increased risk of foodborne listeriosis is expected to occur in the near future.

Listeriosis is an atypical disease with multiple routes of infection, including aerial, cutaneous, transplacental, nosocomial, direct contact, or digestive tract. However, surveillance studies and investigation of recent outbreaks have demonstrated that the most associated transmission pathway to humans is the intake of contaminated food (digestive tract). Ready-to-eat foods, particularly refrigerated foodstuffs, such as milk and dairy products, meat and meat products, raw vegetables, and fruits, have been related to recent outbreaks [3, 4].

The food industry relies on a variety of processing and preservation methods to produce safe and healthy products with adequate shelf life and that are appreciated by consumers. These methods inactivate or inhibit the growth of pathogenic microorganisms such as *L. monocytogenes* and suppress undesirable chemical and biochemical changes, thereby ensuring food safety and maintaining desirable physical and sensory properties. The methods currently used in food preservation involve physical, chemical, or biological factors. In combination with other strategies, refrigeration, freezing, addition of acidifying agents or curing agents (e.g., sodium chloride and sodium nitrite), radiation and high-pressure processing are the most reliable and used preservation techniques. However, there are studies which demonstrate that *L. monocytogenes* strains have mechanisms that allow them to survive and resist the stresses caused by these processing methods [5].

This review focuses on key issues such as the molecular mechanisms underlying *L. monocytogenes* survival and adaptation to stresses caused by different environmental conditions. Since many of the stresses can be found in both food and humans, we will try to correlate these molecular mechanisms with the organism's virulence. Studies on the development of technologies to control and prevent the contamination of *L. monocytogenes* in food matrices and food processing facilities are also briefly discussed.

2. Cold stress response

Cold stress adaptation is a fundamental characteristic of *L. monocytogenes* that markedly contributes to the microorganisms' dissemination via refrigerated food products. Although most foodborne pathogens are effectively controlled under cooling storage, *L. monocytogenes* proliferation persists so, cold-stored contaminated foods provide proper conditions for survival and growth of these organisms [6, 7].

L. monocytogenes, as a psychrotolerant bacterium, is able to grow over a wide range of temperatures (1–45°C), although the optimum temperature range is from 30 to 37°C [8]. Cold stress adaptation in *L. monocytogenes* is mediated through many molecular response mechanisms whose nature remains mainly vague, besides some aspects of this phenomenon have been clarified in model microorganisms.

2.1. Listerial mechanisms of low-temperature resistance

L. monocytogenes response to cold shock comprises the synthesis of cold-shock proteins (CSPs), while during balanced growth at low temperatures, it produces cold acclimation proteins (CAPs). Twelve CSPs and four CAPs were identified as a result of cold stress [9]. The main functions involving CSPs include chaperones involved in DNA recombination course, transcription, translation, and protein folding [10]. The cold adaptation of this pathogen is accompanied by gene expression changes. When cultured at 10°C, *L. monocytogenes* RNAs are increasingly synthesized compared to growth at 37°C [11]. A higher mRNA expression for chaperone proteases suggests that ClpP, ClpB, and GroEL enzymes may participate in the degradation of damaged or abnormal polypeptides arising due to growth at low temperatures.

Changes in temperature also lead to an alteration in the membrane lipid composition to maintain the ideal membrane fluidity required for proper enzyme activity and transport of solutes [12]. *Listeria* cell membrane contains high amounts of iso and anteiso, odd-numbered, branched-chain fatty acids (>95%). When grown under refrigeration temperatures, the anteiso-C15:0 represents 65–85% of total membrane fatty acids. When grown at 37°C, predominant fatty acids are anteiso-C15:0 (41–52%), anteiso-C17:0 (24–51%), and iso-C15:0 (2–18%) [13]. Growth at low temperatures also causes an increase of unsaturated fatty acids, which helps enhancing the fluidity of the membrane. Decreasing the growth temperature from 20 to 5°C precedes a switch from iso to anteiso branching (i-C15:0 to a-C15:0) and a fatty acid shortening (a decrease in C17:0). Annous et al. [13] suggested that the growth of *L. monocytogenes* in refrigerated foods could be controlled by food-grade agents inhibiting the biosynthesis of anteiso-C15:0.

L. monocytogenes growth at low temperatures is also stimulated by the presence of cryoprotectant compatible solutes, for example, betaine, glycine, and carnitine [14, 15]. *Listeria* imports and accumulates these solutes from the environment, and this is one of the functions of sigma factor σ^B (*Listeria*'s general stress transcription factor) during growth at low temperature [16]. In response to cold shock, σ^B controls the transcription of genes encoding the BetL, Gbu, and OpuC uptake system, involved in the accumulation of glycine, betaine, and carnitine. Studies with mutants having deleted osmolyte transporter genes demonstrated the cryoprotective activity of these compounds [17].

3. Acid stress response

L. monocytogenes may be exposed to high acidity levels while in the food chain and during gastrointestinal (GI) passage in the host (i.e., following exposure to fatty acids, in the phagosome of macrophages during systemic infection, and even upon exiting the host, due to fluctuations in environmental pH).

Being a neutrophile (optimum pH 6 or 7), *L. monocytogenes* keeps the intracytoplasmic pH close to neutrality, though pH oscillations in the external medium are imperative for its survival and a prerequisite for pathogenesis and infection [18]. Acid tolerance response (ATR) is the adaptive phenomenon that permits the pathogen to preserve pH homeostasis when exposed to low pH. Understanding the molecular mechanisms of acid adaptation and pH homeostasis is essential in order to control the pathogen growth in high-risk foods and predict the ability to cause disease.

3.1. Listerial mechanisms of acid resistance

Cellular exposure to pH stress induces the modulation of fatty acid profiles in *Listeria* cell membrane, although the changes differ from those documented for other genera [19]. In *L. monocytogenes*, larger proportions of linear chain fatty acids are incorporated into the membrane, with increased levels of C14:0 and C16:0 and a reported concomitant decrease in C18:0 [20, 21].

Under high acidic environments, two chaperonins (DnaK and GroES) and a serine protease (HtrA) have been identified and characterized in *Listeria*, being necessary for the organism survival [22–24]. Other studies shed light on the role of σ^B in modulating genes involved in pH homeostasis and gastrointestinal persistence, thus crucial in *L. monocytogenes* survival after exposure to acid conditions. It has been reported that *Listeria* mutants that lack a *sigB* functional gene exhibit a decreased resistance to low pH conditions, besides σ^B regulates the expression of OpuC, a cold-activated transporter for carnitine.

Additional mechanisms of acid resistance such as the F0-ATPase complex, arginine deiminase system (ADI), and the glutamate decarboxylase (GAD) have been elucidated.

3.1.1. F0F1-ATPase complex

F_0F_1-ATPase is an enzyme organized in two distinct although physically linked domains. The catalytic part (F_1) is cytoplasmic while the integral membrane domain (F_0) acts as a membrane channel for proton translocation. Cytoplasmic domain may either catalyze the synthesis of adenosine triphosphate (ATP) when the protons pass into the cytoplasm through the membrane-bound domain, or hydrolyze ATP when the protons move outside of the cell. Thus, the F_0F_1-ATPase complex is responsible for the aerobic synthesis of ATP, as a result of protons moving into the cell, and generates a proton motive force anaerobically by expelling protons. As a consequence of the latter mechanism, F_0F_1-ATPase is thought to increase intracellular pH in acidic situations [25].

3.1.2. Arginine deiminase system

This system comprises three enzymes: arginine deiminase (encoded by *arcA*) which catalyzes the hydrolysis of arginine to citrulline and ammonia; ornithine carbamoyltransferase (encoded by *arcB*) which is responsible for converting citrulline to ornithine and carbamoylphosphate, in the presence of phosphate; and carbamate kinase (encoded by *arcC*) which synthesizes ATP from carbamoylphosphate and adenosine diphosphate (ADP).

Arginine is transported into the cell in exchange for an ornithine molecule that is moved outside through the transporter encoded by *arcD*, while the pathway enzymes ultimately catabolize arginine to ornithine, ammonia, and CO_2. Ammonia is produced through the catabolization of arginine via the ADI system combined with intracellular protons to produce ammonium ions. This reaction increases intracellular pH, thus allowing survival in hostile environments that would otherwise be lethal to the cell [26]. In addition, ATP is generated by the system and this can be used for driving out protons through F_1F_0-ATPase [27].

3.1.3. Glutamate decarboxylase system

The GAD enzyme, generally encoded by *gadA* or *gadB*, irreversibly decarboxylates glutamate, producing the neutral γ-aminobutyrate (GABA). This reaction results in an increase of the cytoplasmic pH due to the consumption of an intracellular proton. GABA produced by the decarboxylation reaction is subsequently exchanged on the cell membrane for a glutamate molecule by a glutamate: GABA antiporter, generally encoded by the *gadC* gene [28].

The GAD system is crucial for *L. monocytogenes* acid adaptation and, consequently, for a successful passage through the gastric environment, a necessary condition for latter invasion of intestinal epithelial cells [29]. The loss of genes encoding a GAD enzyme and a glutamate transporter decreases the cell's ability to survive in low pH environments and consequently to cause infection [30]. Stress factors commonly associated with the GI tract (low pH, anaerobiosis, hypo- and hyperosmotic shock, bile salts, and chloride ions) have been shown to induce GAD system expression in a variety of bacteria [31, 32].

4. Osmotic stress response

Osmotic stress defines the osmotic strength variation of an organism environment, which results from desiccation or from a high content of osmotically active compounds (salt or sugars) in the environment, lowering its water activity (a_w). Since the bacterial cytoplasmic membrane is permeable to water but not to most other metabolites, hyper- or hypo-osmotic shock causes an efflux or influx of water, accompanied by a concomitant decrease or an increase in intracellular volume, respectively. In general, the internal osmotic pressure is higher than that of the surrounding medium, generating turgor, the driving force for cell extension, growth, and division. Therefore, the bacterial maintenance of pressure turgor is critical to survival in osmotic stress conditions.

The maximum NaCl concentration that permits *L. monocytogenes* growth ranges from 7 to 10% [33]. This osmotolerance is vital during its infectious cycle, since *L. monocytogenes* encounters elevated osmolarity in the food processing industry and in the gastrointestinal lumen of the host. The response of microorganisms to osmotic stress is called osmoadaptation and holds physiological changes and variations in gene expression patterns [34].

4.1. Listerial mechanisms of osmotic resistance

Compatible solute osmoadaptation is a biphasic response in which elevated levels of potassium cation K^+ (and glutamate, its counter-ion) represent a primary response, succeeded by a significant increase in cytoplasmic concentration of compatible solutes. Cells absorb osmolytes from the external environment to restore osmotic balance within cells. The solute-mediated osmoprotection stimulates the growth of cells subjected to high salt concentrations. Deletions of these osmolyte transporters reduce the growth of *Listeria* under conditions of hyperosmolarity [14, 30, 35]. In addition to previously mentioned compatible solutes (glycine, betaine, and carnitine), proline is important for the survival under hyperosmolarity conditions [36]. σ^B factor, as an important part of the overall stress response of *L. monocytogenes,* mediates the expression of *ctc* gene and the use of betaine and carnitine as osmoprotectors.

In response to osmotic stress, two genes involved in cell envelope modification have been identified: *lmo2085,* a putative peptidoglycan-linked protein, and *lmo1078,* a putative UDP-glucose phosphorylase that catalyzes the formation of UDP-glucose, a precursor of membrane glycolipids and of the cell wall [37].

A further mechanism of osmotic adaptation is the modification of genetic expression leading to an increased or a decreased synthesis of several proteins. Salt-shock proteins are rapidly induced and overexpressed for a short time period, being similar to those induced in cold-shock response (CSPs and CAPs). Among CSPs induced in *L. monocytogenes,* there are two general stress response proteins, DnaK that acts as a heat-shock protein stabilizing cellular proteins and Ctc that is involved in high osmolarity resistance in the lack of osmoprotectants, such as glycine, betaine, and carnitine, in the medium [38]. Additional stress response proteins, including ClpC (an ATPase), ClpP (a protease), and HtrA (a protease), are essential for osmotic and acid stress adaptation in *L. monocytogenes* [39]. HtrA may play a role in degrading misfolded proteins and is beneath LisRK control, a two-component regulatory system important for osmoregulation [36].

5. Desiccation stress response

Desiccation tolerance defines the bacteria's aptitude to survive for extended periods on a surface, deficient of nutrients and water. As so, *L. monocytogenes* desiccation tolerance is most likely associated with the ability to persist in food production surfaces and consequently cross-contaminate food products [40]. The low a_w resulting from high osmolarity decreases turgor pressure in a bacterial cell inhibiting bacterial growth [41]. Drying and addition of salt or sugar are traditional methods to lower food a_w and therefore enhance its prolonged shelf life. *L. monocytogenes* grows optimally at $a_w \geq 0.97$, although it may survive in foods with low a_w [42]. When compared to other common infectious foodborne pathogens, *L. monocytogenes* does not appear to grow at $a_w < 0.90$ but it can survive in these conditions, particularly under refrigeration, for long periods. To date, existing information regarding *L. monocytogenes* desiccation survival is limited and primarily focuses on factors influencing the survival to osmotic stress [40, 43–46]. Strains of serotypes 1/2c and 1/2b were the most tolerant to desiccation, followed by 4b and 1/2a [47]. Hansen and Vogel [46] showed the protective effect of osmoadaptation and also the formation of biofilms on the desiccation survival.

6. High hydrostatic pressure

A high hydrostatic pressure (HHP) represents the application of pressure in the range of 50–1000 MPa, though the inactivation of vegetative cells of bacterial species is typically reached from 300 to 700 MPa, and bacterial spores inactivation demands higher pressure levels up to 1000 MPa [48]. However, depending on the pressure level, HHP treatments can fully inactivate bacteria or impose sublethal injuries. For pressures up to 400 MPa, the integrity of Gram-positive bacterial cells and metabolic activity are maintained, with very limited cell destruction [49]. Over the last years, it has been stated that *L. monocytogenes* is potentially capable of recovering culturability following HHP exposure [49–52]. Physiological studies have also demonstrated that increasing pressure levels results in an accelerated decline of metabolic indicators, such as the activity of the LmrP membrane transport system [53]. These findings suggest that bacteria exposed to HHP are unable to grow due to cell injury, but yet can mount a nonspecific response to high pressure. A proportion of the cell population is able to maintain cellular activity of some kind after HHP, demonstrating the capacity to cellular repair and regrow, when adequate conditions are available [49].

To date, little research has been conducted regarding the mechanisms of bacterial adaptation and resistance to high pressure. Wemekamp-Kamphuis et al. [54] demonstrated that one of the responses that enable *Listeria* survival upon HHP treatment results from induction of the general stress response mediated by σ^B. *L. monocytogenes sigB* deletion mutant was more susceptible to HHP exposure than the wild type, while induction of σ^B resulted in an increased HHP protection relative to the untreated control strain.

Several pressure-induced proteins have been increasingly synthesized when compared to the synthesis of other control proteins at atmospheric pressure [55]. *L. monocytogenes* has shown to actively express many genes as a response to high pressure, but some functional categories appear more affected than others. Genes that tend to be expressed at higher levels under high pressure are genes encoding for transport and binding, signal transduction and chemotaxis, cellular processes, transcriptional regulators, metabolism, and protein fate [56]. The stabilization and maintenance of the bacteria cell is at high focus, showed by the significant regulation of ribosomes and proteins, together with components involved in the cell envelope and the septal ring. It is assumed that the activation of genes involved in the lipid and peptidoglycan biosynthetic pathways is connected to this function. Upregulation of genes associated with generalized repair and maintenance has been proved, where the activation of cold- and heat-shock genes is an example for this [57, 58]. When high pressure demands more energy to be used on repair, energy production and conversion is suppressed. The repression of several energy production/conversion, carbohydrate, and other carbon compound catabolic genes may represent a diminishment of catabolism in cells imposed by HPP treatments. This can be seen by the pressure-induced switch from active growth to a cell repair state, the stationary phase, resulting in a decreased growth rate [59].

Several genes associated with cell formation and shape, as well as synthesis or reassembly of cell-wall constituents, in particular peptidoglycan and fatty acids, were observed to have an increased expression. Because of this, genes involved in such functions can be considered as very central in the response to high pressure. It is presumed that *L. monocytogenes* increases both cell division and cell-envelope-associated gene expression aiming to replace damaged components and thus compensate membrane and wall damages [59].

Cell membranes damage by HPP may possibly be a main cause of inactivation or death in Gram-negative bacteria, but it is fallacious to admit that in Gram-positive bacteria. Cell membrane and wall stabilization in the stationary growth phase do provide a protective effect against HPP, being a major factor for the survival of HPP-induced damage [60]. Beyond cell envelope damage, HPP interferes within the nascent septal ring formation along with other associated cell-wall formation and chromosome segregation processes [59].

7. Stress impact on *L. monocytogenes* virulence

L. monocytogenes has a profound ability to adapt to unfavorable stressful environments, switching from a saprophyte to an intracellular pathogen capable of causing serious infection to the host [61]. In this transformation, σ^B dominates both in the external environment and during gastrointestinal transit, while positive regulatory factor A (PrfA) plays a central role on the intracellular infection. In concert with PrfA, σ^B activates the transcription of several *L. monocytogenes* virulence genes: (1) *bsh*, encoding bile salt hydrolase, essential in gastrointestinal colonization prior to invasion; (2) *inlA*, encoding internalin A, mediates entry into human intestinal epithelial cells; and (3) *gadA*, encoding part of the glutamate decarboxylase system, crucial for acid survival [62]. σ^B also contributes to the transcriptional activation of *prfA*, encoding PrfA, a central virulence regulator of virulence gene expression in *L. monocytogenes* [63].

PrfA-dependent virulence gene cluster or LIPI-1 (*Listeria* pathogenicity island 1) encodes most virulence factors involved in the pathogenic infectious cycle. This chromosomal locus comprises the following genes: (1) *hly*, encoding listeriolysin O (LLO), a pore-forming toxin crucial in the escape from phagocytic vacuoles; (2) *plcA* and *plcB*, encoding two phospholipases C which cooperate with LLO in the escape from bacterial phagosomes; (3) *mpl*, encoding a metalloprotease implicated in the maturation of proenzyme pro-PlcB; (4) *actA*, encoding ActA protein involved in the intra- and intercellular motility of the bacteria; and (5) *prfA*, encoding PrfA, a transcriptional activator of LIPI-1 genes [64]. The expression of additional genes dispersed on the chromosome may be PrfA-regulated, as the internalin locus *inlAB* [65], the genes encoding internalins InlA and InlB cell-wall-anchored proteins which induce *Listeria* phagocytosis [66].

Following the complete genome sequencing of several *L. monocytogenes* strains, an increasing number of virulence-related proteins are being identified and their specific involvement during infectious stages deciphered (**Table 1**).

In addition to other factors, the infectious potential of *L. monocytogenes* is conditioned by the environmental conditions prior to host invasion. A correlation between stress response and virulence seems to exist and associates strains having more effective stress response mechanisms to being also more virulent [84]. Early studies by Durst [84] and Wood and Woodbine [85] demonstrated that cold storage may enhance virulence of some strains because the

http://dx.doi.org/10.5772/intechopen.76287

Involvement	Proteins/function	Ref.
Regulation	**PrfA**	[68]
	Positive regulatory factor A, central virulence regulator of virulence gene transcription.	
	SigmaB (σ^B)	[69]
	General stress transcription factor.	
	CtsR	[70]
	Class III stress-response regulator, a transcription repressor.	
	HrcA	[71]
	Heat regulation at controlling inverted repeat of chaperone expression elements. A transcription repressor.	
Attachment and invasion	**InlA**	[65]
	Internalin A, surface protein that mediates entry into cells expressing its receptor, the E-cadherin.	
	InlB	[72]
	Internalin B, surface protein that mediates entry into cells expressing one of the receptors gC1qR, HGF-SF, Met, and the glycosaminoglycanes (GAGs).	
Lysis of vacuoles	**LLO**	[73]
	Listeriolysin O, hemolysin required for vacuole escape by lysis of the phagosome membrane.	
	PC-PLC	[74]
	Phospholipase activated by proteolytic cleavage involving Mpl or by cellular proteases. Required for the lysis of the double-membrane vacuole.	
	Mpl	[75]
	Metalloprotease required for the maturation of PC-PLC.	
Intracellular multiplication	**Hpt**	[76]
	Hexose phosphate transporter required for intracytosolic proliferation.	
Cell-to-cell spread	**ActA**	[77]
	Actin assembly-inducing protein, involved in cell-to-cell spread.	
Environmental stress response and virulence	**HtrA**	[78]
	Serine protease involved in acid and osmotic stress response.	
	Bsh	[79]
	Bile salt hydrolase involved in the intestinal and hepatic phases of listeriosis.	
	ClpC	[80]
	ATPase protein promoting early bacterial escape from the phagosome of macrophages and thus virulence.	
	ClpP	[81]
	Serine protease involved in proteolysis and required for growth under stress condition.	

Involvement	Proteins/function	Ref.
	DnaKJ	[22]
	Chaperone heat-shock proteins encoded by the dnaK operon and required for phagocytosis.	
	GroES, GroEL	[23]
	Chaperone proteins which regulate HrcA posttranscriptionally.	
	GAD	[29]
	The glutamate decarboxylase system, involved in acid stress response.	
	BetL	[82]
	Glycine betaine transport system I, involved in osmotic stress response.	
	Gbu	[15]
	Glycine betaine transport system II, involved in osmotic stress response.	
	OpuC	[83]
	Carnitine transport system, involved in cold and osmotic stress response.	

Table 1. Stress response and virulence-associated proteins in ***Listeria monocytogenes*** (adapted from reference [67]).

pathogen virulence rather increases when grown under refrigeration than at optimal growth temperature. By contrast, virulence gene expression was reported to be downregulated at temperatures below 30°C, besides PrfA is only formed at 37°C [85]. According to Loh et al. [86], the expression of *prfA* is nearly 16-times higher at 37°C compared to that at 30°C, and imperceptible in cells cultivated at 20°C. The specific pathogenicity of LLO can be fully recovered in less than 24 h by incubating refrigerated cells at 37°C [87]. This virulence recovery after heat shock reinforces the importance of eliminating *L. monocytogenes* from minimally processed ready-to-eat foods held at refrigeration temperatures for long periods.

Low pH and high salt content are common factors often found in foods contaminated with *L. monocytogenes* [89]. Even though at these conditions, the growth of most foodborne and spoilage bacteria is restricted, *L. monocytogenes* is capable of surviving and even grow in such environments; long-term adaptation to these sublethal stress conditions results in altered virulence [88].

Conte et al. [31, 89] demonstrated that short-term exposure (1 h) of *L. monocytogenes* to a sublethal acidic environment (pH 5.1) not only increased its invasiveness to the human colon adenocarcinoma cell line Caco-2 but also increased the ability of *L. monocytogenes* to survive and proliferate in macrophage-like cells, suggesting that exposure to a low pH (e.g., in the human stomach) may enhance listerial overall virulence. In addition, LLO excreted by virulent *L. monocytogenes* showed a maximal activity at pH 4.0–5.0. In another study, the exposure of *L. monocytogenes* to acidic shock has induced the transcription of two important virulence genes (*inlA* and *bsh*) [90]. Conversely, a study by Rieu et al. [91] reported a decrease in virulence gene transcription after 5 h at pH 4.0 achieved with acetic acid. This conflicting finding may be sustained by the use of organic acids since they might be more harmful to the bacteria. Some weak organic acids enhance pathogenicity of the bacterium, while others reduce it, as the secretion of LLO is increased by citrate, acetate, and lactate, whereas sorbate inhibited this hemolysin [92]. This knowledge would be important for the selection of acidulants to be used in different foods.

Garner et al. [93] reported an intensified invasiveness of *L. monocytogenes* for Caco-2 cells when grown at 7°C rather than at 37°C, and, for both temperatures, the invasion ability was greater in cells grown at pH 7.4 compared to growth at pH 5.5. A growth temperature of 37°C, pH 7.4, in the presence of NaCl or sodium lactate, enhanced *L. monocytogenes* invasiveness; however, the pre-exposure to gastric fluid (pH 4.5), even for as short as 10 s, substantially reduced its invasion. These findings intimate that listerial virulence-associated characteristics seem to be affected by specific food properties (e.g., the presence of organic acids or salt). The authors further showed that *L. monocytogenes* growth phase affects its ability to invade Caco-2 cells. The invasion by log-phase cells was 9.5-fold lower than invasion by stationary-phase cells, corroborating other studies which demonstrate that exposure of *L. monocytogenes* to different environmental conditions can change invasiveness and virulence [93]. Accordingly, the increased stationary-phase invasiveness also coincides with stationary-phase induction of σ^B activity [90]. In stationary-phase cells, *inlA* expression is regulated in a σ^B-dependent manner, and growth phase-dependent effects on invasion appear independent of PrfA [94, 95], contributing to *inlA* transcription [96].

Complementary studies demonstrate that *L. monocytogenes* pathogenicity requires an adaptive acid tolerance response, so the ability to survive gastric acid fluid and to invade host cells is related to ATR activation [30, 89, 97]. This finding is supported by the fact that the glutamate decarboxylase (GAD) system, as the ATR most important component, is required for listerial survival in the gastric environment, and also LisRK deletion, a two-component system involved in acid resistance regulation, caused a dramatic reduction in virulence [29, 98].

A further prerequisite for *L. monocytogenes* infection depends on the ability to counteract conditions of elevated osmolarity in the gastrointestinal tract. As mentioned in Section 2.1, the carnitine uptake system (OpuC) is directly linked to osmotic stress resistance of *L. monocytogenes* and to its ability to reach and proliferate in the liver and spleen [17]. Carnitine (produced from the desquamation of the gastrointestinal epithelial layer) was formerly proved to act as a crucial osmoprotectant, facilitating growth in this gastrointestinal environment, once changing the carnitine transported OpuC resulted in a significant reduction in *Listeria* ability to colonize the upper small intestine and cause subsequent systemic infection [99, 100]. A supporting study by Wemekamp-Kamphuis et al. [17] demonstrated that a triple mutant, defective in all three compatible uptake systems (BetL, Gbu, and OpuC), showed a similar phenotype to that of a single opuC mutant, mutually revealing a decreased ability to cause systemic infection relative to the parent. Those were clear evidences that *betL* and *gbu* do not play a significant role in *L. monocytogenes* pathogenesis and that it is the carnitine uptake system that most induces listerial virulence. In addition, Joseph et al. [101] also identified OpuCA and OpuCB as being induced intracellularly. Since the contribution of each transporter is dependent on the external environment, there are occurrences when each system is tailored for optimal effects within a certain environmental niche.

Over the last years, novel trends in food production tend to preserve the natural flavor and texture of products using minimal processing. Non-thermal food preservation usually allows a significant microbial reduction, and mounting evidence also demonstrates that the conditions applied by alternative technologies may influence bacterial virulence [102]. The application of HHP has been shown not to induce mutations in the internal genes, *inlA* and *inlB*, implicated in the adhesion and internalization of *L. monocytogenes* in human cells. However, when the effect of HPP on the *ctsR* gene is observed, a reduction in virulence potential of surviving cells was noted. Likewise, virulence and reduced motility may be the result of a mutation in this gene corresponding to the loss of a single amino acid. This suppression could be related to a high-pressure tolerance [70, 103].

8. Conclusions

Exposure of *L. monocytogenes* to sublethal environmental stresses can enhance its survival to subsequent lethal conditions and additionally induce the expression of the organism's virulence genes. Therefore, exposure of *L. monocytogenes* to food-associated stresses such as high salt concentrations or low temperatures during refrigerated storage may result in increased virulence and thus a higher risk for listeriosis. Any strain of *L. monocytogenes* present in food is actually considered equally pathogenic. However, results from several studies support the idea that the heterogeneity among strains regarding the response to stress and virulence potential should be considered, once responses to food matrix and storage conditions are often strain specific.

Although significant advances in our understanding on stress response and virulence potential have been achieved in the last years, there is still a need to fulfill knowledge gaps on molecular mechanisms behind *L. monocytogenes* response to stress and virulence. Further studies on the influence of food matrix on stress tolerance and virulence potential of different strains, recovered from foods and from patients, are needed. This information can be further used by regulators to refine previous risk assessments and also in the definition of control measures by the food industry.

Acknowledgements

This work was supported by National Funds from FCT—Fundação para a Ciência e a Tecnologia through project UID/Multi/50016/2013. Publication in open access was co-financed by the project NORTE-01-0246-FEDER-000011, supported by Norte Portugal Regional Operational Programme (NORTE 2020), under the PORTUGAL 2020 Partnership Agreement, through the European Regional Development Fund (ERDF). Financial support for author Sofia Pereira was provided by ESF—European Social Fund, under the PORTUGAL 2020 Partnership Agreement, through doctoral fellowship NORTE-08-5369-FSE-000007_BD_1.

Author details

Sofia Araújo Pereira, Ângela Alves, Vânia Ferreira and Paula Cristina Maia Teixeira*

*Address all correspondence to: pcteixeira@porto.ucp.pt

Universidade Católica Portuguesa, CBQF - Centro de Biotecnologia e Química Fina – Laboratório Associado, Escola Superior de Biotecnologia, Porto, Portugal

References

[1] Saavedra L, Bellomio A, Hebert EM, Minahk C, Suarez N, Sesma F. Listeria: Epidemiology, pathogenesis and novel potential treatments. In: Romano A, Giordano CF, editors. Listeria Infections: Epidemiology, Pathogenesis and Treatment. New York: Nova Science Publishers, Incorporated; 2012. pp. 67-98

[2] EFSA. The European Union summary report on trends and sources of zoonoses, zoonotic agents and food-borne outbreaks in 2016. EFSA Journal. 2017;**15**(12):56-78

[3] Magalhães R, Almeida G, Ferreira V, Santos I, Silva J, Mendes MM, Pita J, Mariano G, Mâncio I, Sousa MM, Farber J, Pagotto F, Teixeira P. Cheese-related listeriosis outbreak, Portugal, March 2009 to February 2012. Euro Surveillaince. 2015;**20**(17):pii=21104

[4] Callejón RM, Rodríguez-Naranjo MI, Ubeda C, Hornedo-Ortega R, Garcia-Parrilla MC, Troncoso AM. Reported foodborne outbreaks due to fresh produce in the United States and European Union: Trends and causes. Foodborne Pathogens and Disease. 2015 Jan;**12**(1):32-38

[5] Lou Y, Yousef AE. Adaptation to sublethal environmental stresses protects *Listeria monocytogenes* against lethal preservation factors. Applied and Environmental Microbiology. 1997;**63**(4):1252-1255

[6] Junttila JR, Niemelä SI, Hirn J. Minimum growth temperatures of *Listeria monocytogenes* and non-haemolytic Listeria. The Journal of Applied Bacteriology. 1988 Oct;**65**(4):321-327

[7] Walker SJ, Archer P, Banks JG. Growth of *Listeria monocytogenes* at refrigeration temperatures. The Journal of Applied Bacteriology. 1990 Feb;**68**(2):157-162

[8] Seeliger HPR, Jones D. Listeria. In: Sneath PHA, Mair NS, Sharpe NE, Holt JG, editors. Bergey's Manual of Systematic Bacteriology. Vol 2. Baltimore: Williams and Wilkins; 1986. pp. 1235-1245

[9] Bayles DO, Annous BA, Wilkinson BJ. Cold stress proteins induced in *Listeria monocytogenes* in response to temperature downshock and growth at low temperatures. Applied and Environmental Microbiology. 1996 Mar;**62**(3):1116-9

[10] Schmid B, Klumpp J, Raimann E, Loessner MJ, Stephan R, Tasara T. Role of cold shock proteins in growth of *Listeria monocytogenes* under cold and osmotic stress conditions. Applied and Environmental Microbiology. 2009 Mar;**75**(6):1621-1627

[11] Liu S, Graham JE, Bigelow L, Ii PDM, Wilkinson BJ. Identification of *Listeria monocytogenes* genes expressed in response to growth at low temperature. Applied and Environmental Microbiology. 2002;**68**(4):1697-1705

[12] Mansilla MC, Cybulski LE, Albanesi D, de Mendoza D. Control of membrane lipid fluidity by molecular thermosensors. Journal of Bacteriology. 2004 Oct;**186**(20):6681-6688

[13] Annous BA, Becker LA, Bayles DO, Labeda DP, Wilkinson BJ. Critical role of anteiso-C15:0 fatty acid in the growth of *Listeria monocytogenes* at low temperatures. Applied and Environmental Microbiology. 1997 Oct;**63**(10):3887-3894

[14] Angelidis AS, Smith GM. Role of the glycine betaine and carnitine transporters in adaptation of *Listeria monocytogenes* to chill stress in defined medium. Applied and Environmental Microbiology. 2003 Dec;**69**(12):7492-7498

[15] Lou MM, Smith LT. Gbu glycine betaine porter and carnitine uptake in osmotically stressed *Listeria monocytogenes* cells. Applied and Environmental Microbiology. 2002 Nov;**68**(11):5647-5655

[16] Bayles DO, Wilkinson BJ. Osmoprotectants and cryoprotectants for *Listeria monocytogenes*. Letters in Applied Microbiology. 2000 Jan;**30**(1):23-27

[17] Wemekamp-Kamphuis HH, Wouters JA, Sleator RD, Gahan CGM, Hill C, Abee T. Multiple deletions of the osmolyte transporters BetL, Gbu, and OpuC of *Listeria monocytogenes* affect virulence and growth at high osmolarity. Applied and Environmental Microbiology. 2002 Oct;**68**(10):4710-4716

[18] Bearson S, Bearson B, Foster JW. Acid stress responses in enterobacteria. FEMS Microbiology Letters. 1997 Feb;**147**(2):173-180

[19] Fozo EM, Quivey RG. The fabM gene product of streptococcus mutans is responsible for the synthesis of monounsaturated fatty acids and is necessary for survival at low pH. Journal of Bacteriology. 2004;**186**(13):4152-4158

[20] van Schaik W, Gahan CG, Hill C. Acid-adapted *Listeria monocytogenes* displays enhanced tolerance against the lantibiotics nisin and lacticin 3147. Journal of Food Protection. 1999 May;**62**(5):536-539

[21] Ryan S, Hill C, Gahan CGM. Acid stress responses in *Listeria monocytogenes*. Advances in Applied Microbiology. 2008;**65**:67-91

[22] Hanawa T, Fukuda M, Kawakami H, Hirano H, Kamiya S, Yamamoto T. The *Listeria monocytogenes* DnaK chaperone is required for stress tolerance and efficient phagocytosis with macrophages. Cell Stress Chaperones. 1999 Jun;**4**(2):118-128

[23] Gahan CGM, O'Mahony J, Hill C. Characterization of the groESL operon in *Listeria monocytogenes*: Utilization of two reporter systems (gfp and hly) for evaluating in vivo expression. Infection and Immunity. 2001 Jun;**69**(6):3924-3932

[24] Wilson RL, Brown LL, Kirkwood-Watts D, Warren TK, Lund SA, King DS, et al. *Listeria monocytogenes* 10403S HtrA is necessary for resistance to cellular stress and virulence. Infection and Immunity. 2006 Jan;**74**(1):765-768

[25] Cotter PD, Gahan CG, Hill C. Analysis of the role of the *listeria monocytogenes* F0F1-AtPase operon in the acid tolerance response. International Journal of Food Microbiology. 2000 Sep;**60**(2-3):137-146

[26] Ryan S, Begley M, Gahan CGM, Hill C. Molecular characterization of the arginine deiminase system in *Listeria monocytogenes*: Regulation and role in acid tolerance. Environmental Microbiology. 2009 Feb;**11**(2):432-445

[27] Lucas PM, Blancato VS, Claisse O, Magni C, Lolkema JS, Lonvaud-Funel A. Agmatine deiminase pathway genes in lactobacillus brevis are linked to the tyrosine decarboxylation operon in a putative acid resistance locus. Microbiology. 2007 Jul;**153**(7):2221-2230

[28] Karatzas K-AG, Suur L, O'Byrne CP. Characterization of the intracellular glutamate decarboxylase system: Analysis of its function, transcription, and role in the acid resistance of various strains of *Listeria monocytogenes*. Applied and Environmental Microbiology. 2012 May;**78**(10):3571-3579

[29] Cotter PD, Gahan CG, Hill C. A glutamate decarboxylase system protects *Listeria monocytogenes* in gastric fluid. Molecular Microbiology. 2001 Apr;**40**(2):465-475

[30] Gandhi M, Chikindas ML. Listeria: A foodborne pathogen that knows how to survive. International Journal of Food Microbiology. 2007 Jan;**113**(1):1-15

[31] Conte MP, Petrone G, Di Biase AM, Longhi C, Penta M, Tinari A, et al. Effect of acid adaptation on the fate of *Listeria monocytogenes* in THP-1 human macrophages activated by gamma interferon. Infection and Immunity. 2002 Aug;**70**(8):4369-4378

[32] Jydegaard-Axelsen A-M, Hoiby PE, Holmstrom K, Russell N, Knochel S. CO2- and anaerobiosis-induced changes in physiology and gene expression of different *Listeria monocytogenes* strains. Applied and Environmental Microbiology. 2004 Jul;**70**(7):4111-4117

[33] Lado BH, Yousef AE. Characteristics of *Listeria monocytogenes* important to food processors. In: Ryser T, Marth EH, editors. Listeria, Listeriosis and Food Safety. Boca Raton, FL: CRC Press; 2007. pp. 157-214

[34] Hill C, Cotter PD, Sleator RD, Gahan CGM. Bacterial stress response in Listeria monocytogenes: Jumping the hurdles imposed by minimal processing. International Dairy Journal. 2002;**12**:273-283

[35] Wemekamp-kamphuis HH, Sleator RD, Wouters JA, Hill C, Abee T. Molecular and physiological analysis of the role of osmolyte transporters BetL, Gbu, and OpuC in growth of *Listeria monocytogenes* at low temperatures. Applied and Environmental Microbiology. 2004;**70**(5):2912-2918

[36] Sleator RD, Hill C. Bacterial osmoadaptation: The role of osmolytes in bacterial stress and virulence. FEMS Microbiology Reviews. 2002 Mar;**26**(1):49-71

[37] Utratna M, Shaw I, Starr E, O'Byrne CP. Rapid, transient, and proportional activation of σ(B) in response to osmotic stress in *Listeria monocytogenes*. Applied and Environmental Microbiology. 2011 Nov;**77**(21):7841-7845

[38] Duché O, Trémoulet F, Glaser P, Labadie J. Salt stress proteins induced in *Listeria monocytogenes*. Applied and Environmental Microbiology. 2002 Apr;**68**(4):1491-1498

[39] Wonderling LD, Wilkinson BJ, Bayles DO. The htrA (degP) gene of *Listeria monocytogenes* 10403S is essential for optimal growth under stress conditions. Applied and Environmental Microbiology. 2004 Apr;**70**(4):1935-1943

[40] Vogel BF, Hansen LT, Mordhorst H, Gram L. The survival of *Listeria monocytogenes* during long term desiccation is facilitated by sodium chloride and organic material. International Journal of Food Microbiology. 2010 Jun 15;**140**(2-3):192-200

[41] Amezaga M, Davidson I, McLaggan D, Verheul A, Abee T, Booth I. The role of peptide metabolism in the growth of *Listeria monocytogenes* ATCC 23074 at high osmolarity. Microbiology. 1995 Jan;**141**(1):41-49

[42] Nolan DA, Chamblin DC, Troller JA. Minimal water activity levels for growth and survival of *Listeria monocytogenes* and *Listeria innocua*. International Journal of Food Microbiology. 1992 Aug;**16**(4):323-335

[43] Zoz F, Iaconelli C, Lang E, Iddir H, Guyot S, Grandvalet C, et al. Control of relative air humidity as a potential means to improve hygiene on surfaces: A preliminary approach with *Listeria monocytogenes*. Almeida A, editor. PLoS One. 2016 Feb;**11**(2):e0148418

[44] Overney A, Chassaing D, Carpentier B, Guillier L, Firmesse O. Development of synthetic media mimicking food soils to study the behaviour of *listeria monocytogenes* on stainless steel surfaces. International Journal of Food Microbiology. 2016 Dec;**238**:7-14

[45] Hingston PA, Piercey MJ, Hansen LT. Genes associated with desiccation and osmotic stress in *Listeria monocytogenes* as revealed by insertional mutagenesis. Applied and Environmental Microbiology. 2015 Aug;**81**(16):5350-5362

[46] Hansen LT, Vogel BF. Desiccation of adhering and biofilm *Listeria monocytogenes* on stainless steel: Survival and transfer to salmon products. International Journal of Food Microbiology. 2011 Mar;**146**(1):88-93

[47] Hingston P, Chen J, Dhillon BK, Laing C, Bertelli C, Gannon V, et al. Genotypes associated with *Listeria monocytogenes* isolates displaying impaired or enhanced tolerances to cold, salt, acid, or desiccation stress. Frontiers in Microbiology. 2017 Mar;**8**:369

[48] Smelt JPP. Recent advances in the microbiology of high pressure processing. Trends in Food Science & Technology. 1998 Apr;**9**(4):152-158

[49] Ritz M, Pilet MF, Jugiau F, Rama F, Federighi M. Inactivation of salmonella Typhimurium and *listeria monocytogenes* using high-pressure treatments: Destruction or sublethal stress? Letters in Applied Microbiology. 2006;**42**(4):357-362

[50] Jantzen MM, Navas J, de Paz M, Rodriguez B, da Silva WP, Nunez M, et al. Evaluation of ALOA plating medium for its suitability to recover high pressure-injured *Listeria monocytogenes* from ground chicken meat. Letters in Applied Microbiology. 2006 Sep;**43**(3):313-317

[51] Bozoglu F, Alpas H, Kaletunç G. Injury recovery of foodborne pathogens in high hydrostatic pressure treated milk during storage. FEMS Immunology and Medical Microbiology. 2004 Apr;**40**(3):243-247

[52] Bull MK, Hayman MM, Stewart CM, Szabo EA, Knabel SJ. Effect of prior growth temperature, type of enrichment medium, and temperature and time of storage on recovery of *Listeria monocytogenes* following high pressure processing of milk. International Journal of Food Microbiology. 2005;**101**(1):53-61

[53] Kilimann K, Hartmann C, Delgado A, Vogel R, Ganzle M. A fuzzy logic-based model for the multistage high-pressure inactivation of ssp. MG 1363. International Journal of Food Microbiology. 2005 Jan;**98**(1):89-105

[54] Wemekamp-Kamphuis HH, Wouters JA, de Leeuw PPLA, Hain T, Chakraborty T, Abee T. Identification of sigma factor B-controlled genes and their impact on acid stress, high

hydrostatic pressure, and freeze survival in *Listeria monocytogenes* EGD-e. Applied and Environmental Microbiology. 2004 Jun;**70**(6):3457-3466

[55] Welch TJ, Farewell A, Neidhardt FC, Bartlett DH. Stress response of *Escherichia coli* to elevated hydrostatic pressure. Journal of Bacteriology. 1993 Nov;**175**(22):7170-7177

[56] Liu Y, Ream A. Gene expression profiling of *Listeria monocytogenes* strain F2365 during growth in ultrahigh-temperature-processed skim milk. Applied and Environmental Microbiology. 2008 Nov;**74**(22):6859-6866

[57] Malone AS, Chung Y-K, Yousef AE. Genes of *Escherichia coli* O157:H7 that are involved in high-pressure resistance. Applied and Environmental Microbiology. 2006 Apr; **72**(4):2661-2671

[58] Scortti M, Monzó HJ, Lacharme-Lora L, Lewis DA, Vázquez-Boland JA. The PrfA virulence regulon. Microbes and Infection. 2007 Aug;**9**(10):1196-1207

[59] Bowman JP, Bittencourt CR, Ross T. Differential gene expression of *Listeria monocytogenes* during high hydrostatic pressure processing. Microbiology. 2008;**154**(2):462-475

[60] Mañas P, Mackey BM. Morphological and physiological changes induced by high hydrostatic pressure in exponential- and stationary-phase cells of Escherichia coli: Relationship with cell death. Applied and Environmental Microbiology. 2004 Mar;**70**(3):1545-1554

[61] Freitag NE, Port GC, Miner MD. *Listeria monocytogenes*—From saprophyte to intracellular pathogen. Nature Reviews. Microbiology. 2009 Sep;**7**(9):623-628

[62] Kazmierczak MJ, Mithoe SC, Boor KJ, Wiedmann M. *Listeria monocytogenes* sigma B regulates stress response and virulence functions. Journal of Bacteriology. 2003 Oct;**185**(19):5722-5734

[63] Nadon CA, Bowen BM, Wiedmann M, Boor KJ. Sigma B contributes to PrfA-mediated virulence in *Listeria monocytogenes*. Infection and Immunity. 2002;**70**(7):3948

[64] Leimeister-Wächter M, Haffner C, Domann E, Goebel W, Chakraborty T. Identification of a gene that positively regulates expression of listeriolysin, the major virulence factor of *listeria monocytogenes*. Proceedings of the National Academy of Sciences of the United States of America. 1990;**87**(21):8336

[65] Gaillard JL, Berche P, Frehel C, Gouln E, Cossart P, Gouin E, et al. Entry of *L. monocytogenes* into cells is mediated by internalin, a repeat protein reminiscent of surface antigens from gram-positive cocci. Cell. 1991 Jun;**65**(7):1127-1141

[66] Kuhn M, Goebel W. Internalization of *listeria monocytogenes* by nonprofessional and professional phagocytes. In: Oelschlaeger TA, Hacker JH, editors. Bacterial Invasion into Eukaryotic Cells Subcellular Biochemistry. Vol. 33. Boston, MA: Springer; 2000. pp. 411-436

[67] Roche SM, Velge P, Liu D. Virulence determination, section II identification and detection. In: Liu D, editor. Handbook of *Listeria monocytogenes*. Boca Raton: CRC Press, Taylor & Francis Group; 2008. pp. 241-270

[68] Kreft J, Vázquez-Boland JA. Regulation of virulence genes in Listeria. International Journal of Medical Microbiology. 2001;**291**:145-157

[69] Wiedmann M, Arvik TJ, Hurley RJ, Boor KJ. General stress transcription factor sigmaB and its role in acid tolerance and virulence of *Listeria monocytogenes*. Journal of Bacteriology. 1998 Jul;**180**(14):3650-3656

[70] Karatzas KAG, Wouters JA, Gahan CGM, Hill C, Abee T, Bennik MHJ. The CtsR regulator of *Listeria monocytogenes* contains a variant glycine repeat region that affects piezotolerance, stress resistance, motility and virulence. Molecular Microbiology. 2003 Sep;**49**(5):1227-1238

[71] van der Veen S, Abee T. HrcA and DnaK are important for static and continuous-flow biofilm formation and disinfectant resistance in *listeria monocytogenes*. Microbiology. 2010 Dec;**156**(12):3782-3790

[72] Bierne H, Cossart P. InlB, a surface protein of *Listeria monocytogenes* that behaves as an invasin and a growth factor. Journal of Cell Science. 2002 Sep;**115**(Pt 17):3357-3367

[73] Cossart P, Vicente MF, Mengaud J, Baquero F, Perez-Diaz JC, Berche P. Listeriolysin O is essential for virulence of *listeria monocytogenes*: Direct evidence obtained by gene complementation. Infection and Immunity. 1989 Nov;**57**(11):3629-3636

[74] Marquis H, Doshi V, Portnoy DA. The broad-range phospholipase C and a metalloprotease mediate listeriolysin O-independent escape of *Listeria monocytogenes* from a primary vacuole in human epithelial cells. Infection and Immunity. 1995 Nov;**63**(11):4531-4534

[75] Poyart C, Abachin E, Razafimanantsoa I, Berche P. The zinc metalloprotease of *Listeria monocytogenes* is required for maturation of phosphatidylcholine phospholipase C: Direct evidence obtained by gene complementation. Infection and Immunity. 1993 Apr;**61**(4):1576-1580

[76] Chico-Calero I, Suarez M, Gonzalez-Zorn B, Scortti M, Slaghuis J, Goebel W, et al. Hpt, a bacterial homolog of the microsomal glucose-6-phosphate translocase, mediates rapid intracellular proliferation in Listeria. Proceedings of the National Academy of Sciences. 2002 Jan;**99**(1):431-436

[77] Kocks C, Gouin E, Tabouret M, Berche P, Ohayon H, Cossart P. *L. monocytogenes*-induced actin assembly requires the actA gene product, a surface protein. Cell. 1992 Feb;**68**(3):521-531

[78] Stack HM, Sleator RD, Bowers M, Hill C, Gahan CGM. Role for HtrA in stress induction and virulence potential in *Listeria monocytogenes*. Applied and Environmental Microbiology. 2005 Aug;**71**(8):4241-4247

[79] Dussurget O, Cabanes D, Dehoux P, Lecuit M, Buchrieser C, Glaser P, et al. *Listeria monocytogenes* bile salt hydrolase is a PrfA-regulated virulence factor involved in the intestinal and hepatic phases of listeriosis. Molecular Microbiology. 2002 Aug;**45**(4):1095-1106

[80] Rouquette C, de Chastellier C, Nair S, Berche P. The ClpC ATPase of *Listeria monocytogenes* is a general stress protein required for virulence and promoting early bacterial

escape from the phagosome of macrophages. Molecular Microbiology. 1998 Mar;**27**(6):1235-1245

[81] Gaillot O, Pellegrini E, Bregenholt S, Nair S, Berche P. The ClpP serine protease is essential for the intracellular parasitism and virulence of *Listeria monocytogenes*. Molecular Microbiology. 2000 Mar;**35**(6):1286-1294

[82] Sleator RD, Gahan CG, Abee T, Hill C. Identification and disruption of BetL, a secondary glycine betaine transport system linked to the salt tolerance of *Listeria monocytogenes* LO28. Applied and Environmental Microbiology. 1999 May;**65**(5):2078-2083

[83] Angelidis AS, Smith LT, Hoffman LM, Smith GM. Identification of opuC as a chill-activated and osmotically activated carnitine transporter in *Listeria monocytogenes*. Applied and Environmental Microbiology. 2002 Jun;**68**(6):2644-2650

[84] Roche SM, Gracieux P, Milohanic E, Albert I, Virlogeux-Payant I, Temoin S, et al. Investigation of specific substitutions in virulence genes characterizing phenotypic groups of low-virulence field strains of *Listeria monocytogenes*. Applied and Environmental Microbiology. 2005 Oct;**71**(10):6039-6048

[85] Johansson J, Mandin P, Renzoni A, Chiaruttini C, Springer M, Cossart P. An RNA thermosensor controls expression of virulence genes in *Listeria monocytogenes*. Cell. 2002 Sep;**110**(5):551-561

[86] Loh E, Dussurget O, Gripenland J, Vaitkevicius K, Tiensuu T, Mandin P, et al. A trans-acting Riboswitch controls expression of the virulence regulator PrfA in *Listeria monocytogenes*. Cell. 2009 Nov;**139**(4):770-779

[87] Buncic S, Avery SM, Rogers AR. Listeriolysin O production and pathogenicity of non-growing *Listeria monocytogenes* stored at refrigeration temperature. International Journal of Food Microbiology. 1996 Aug;**31**(1-3):133-147

[88] Cole MB, Jones MV, Holyoak C. The effect of pH, salt concentration and temperature on the survival and growth of *Listeria monocytogenes*. The Journal of Applied Bacteriology. 1990 Jul;**69**(1):63-72

[89] Conte M, Petrone G, Di Biase A, Ammendolia M, Superti F, Seganti L. Acid tolerance in *Listeria monocytogenes* influences invasiveness of enterocyte-like cells and macrophage-like cells. Microbial Pathogenesis. 2000 Sep;**29**(3):137-144

[90] Sue D, Fink D, Wiedmann M, Boor KJ. Sigma B-dependent gene induction and expression in *Listeria monocytogenes* during osmotic and acid stress conditions simulating the intestinal environment. Microbiology. 2004 Nov;**150**(11):3843-3855

[91] Rieu A, Guzzo J, Piveteau P. Sensitivity to acetic acid, ability to colonize abiotic surfaces and virulence potential of *Listeria monocytogenes* EGD-e after incubation on parsley leaves. Journal of Applied Microbiology. 2010 Feb;**108**(2):560-570

[92] Kouassi Y, Shelef LA. Listeriolysin O secretion by *Listeria monocytogenes* in broth containing salts of organic acids. Journal of Food Protection. 1995;**58**(12):1314-1319

[93] Garner MR, James KE, Callahan MC, Wiedmann M, Boor KJ. Exposure to salt and organic acids increases the ability of *Listeria monocytogenes* to invade Caco-2 cells but decreases its ability to survive gastric stress. Applied and Environmental Microbiology. 2006;**72**(8):5384-5395

[94] Kim H, Boor KJ, Marquis H. *Listeria monocytogenes* B contributes to invasion of human intestinal epithelial cells. Infection and Immunity. 2004 Dec;**72**(12):7374-7378

[95] Kim H, Marquis H, Boor KJ. B contributes to *Listeria monocytogenes* invasion by controlling expression of inlA and inlB. Microbiology. 2005 Oct;**151**(10):3215-3222

[96] Dramsi S, Kocks C, Forestier C, Cossart P. Internalin-mediated invasion of epithelial cells by *Listeria monocytogenes* is regulated by the bacterial growth state, temperature and the pleiotropic activator prfA. Molecular Microbiology. 1993 Sep;**9**(5):931-941

[97] O'Driscoll B, Gahan CG, Hill C. Adaptive acid tolerance response in *Listeria monocytogenes*: Isolation of an acid-tolerant mutant which demonstrates increased virulence. Applied and Environmental Microbiology. 1996 May;**62**(5):1693-1698

[98] Cotter PD, Emerson N, Gahan CG, Hill C. Identification and disruption of lisRK, a genetic locus encoding a two-component signal transduction system involved in stress tolerance and virulence in *Listeria monocytogenes*. Journal of Bacteriology. 1999 Nov;**181**(21):6840-6843

[99] Sleator RD, Francis GA, O'Beirne D, Gahan CGM, Hill C. Betaine and carnitine uptake systems in *Listeria monocytogenes* affect growth and survival in foods and during infection. Journal of Applied Microbiology. 2003;**95**(4):839-846

[100] Sleator RD, Gahan CGM, O'Driscoll B, Hill C. Analysis of the role of betL in contributing to the growth and survival of *Listeria monocytogenes* LO28. International Journal of Food Microbiology. 2000 Sep;**60**(2-3):261-268

[101] Joseph B, Przybilla K, Stuhler C, Schauer K, Slaghuis J, Fuchs TM, et al. Identification of *Listeria monocytogenes* genes contributing to intracellular replication by expression profiling and mutant screening. Journal of Bacteriology. 2006 Jan;**188**(2):556-568

[102] van Schaik W, Abee T. The role of σB in the stress response of Gram-positive bacteria—Targets for food preservation and safety. Current Opinion in Biotechnology. 2005 Apr;**16**(2):218-224

[103] Van Boeijen IKH, Casey PG, Hill C, Moezelaar R, Zwietering MH, Gahan CGM, et al. Virulence aspects of *Listeria monocytogenes* LO28 high pressure-resistant variants. Microbial Pathogenesis. 2013 Jun;**59-60**:48-51